THE ORIGIN OF LIFE CIRCUS

A How To Make Life Extravaganza

SUZAN MAZUR

For Peter and Florence Rosemary

SUZAN MAZUR

THE ORIGIN OF LIFE CIRCUS

A HOW TO MAKE LIFE EXTRAVAGANZA

This is an extraordinary, timely book about what author Suzan Mazur considers the "greatest story" there is for a journalist to cover: the politics of the investigation into the origin and synthesis of life.
--**Adrian Bejan, NASA Book Review**

CASWELL BOOKS   NEW YORK

Contents

The Origin of Life Circus

Science is about social momentum. It's a circus of controversy, particularly origin of life science, which isn't really even a science -- it's a Big Top of exquisite conjecture, a field of wild imaginings waiting to be tamed inside the Who We Are And Where We Came From tent. No one can yet say what the origin of life narrative is despite the fanfare: trumpets, thundering hooves, flared nostrils, firebreathing and other posturing, and so the clash of "expert" opinion makes for a light show with all the color of a supercollider collision event. . .

Somewhere around the spring of 2012, I began to notice that the investigation into origin of life was ramping up. I'd previously learned during interviews in 2008 for my last book that there were already a thousand scientists in various parts of the world probing life's greatest mystery through an affiliation with NASA Astrobiology Institute's "search for the origin, distribution and future of life in the Universe." Former NAI director Bruce Runnegar told me at the time that there were also hundreds of institutions receiving funding from NAI. Of course that didn't account for those laboratories unrelated to NAI researching origin of life, including synthetic cell development, in whole or in part.

In the past six years, there's also been an increasing cross-disciplinary flavor to the field. Young scientists have been encouraged to pursue a career in origin of life research whether their area is biology, chemistry, geology, physics, biophysics, biochemistry, geochemistry, geophysics, astrophysics, whatever.

Philanthropy, a substantial amount coming indirectly from Wall Street, is currently driving interest in the origin of life investigation. Scientists are also seeking direct investment. A few are even selling origin of life jewelry -- peridot, apatite, a rainbow of tourmaline -- whatever it takes to keep the circus wagons rolling.

It's an irresistible story that few journalists are following, partly because origin of life is a difficult subject. Fewer still have been covering the politics, even though as I write, the battle lines of the origin of life camps, *i.e.*, the metabolists, the compartmentalists, and the geneticists are still clearly visible. So I began reporting on developments and controversies in a series of feature interviews with dozens of the most compelling scientists. Excerpts of those conversations follow.

The beginning of my investigation into the investigation of origin of life roughly coincided with Harry Lonsdale's Origin of Life Challenge, his pledge of $2M for the best papers, the best hypotheses for how life started. There was now a race on!

Harry Lonsdale considered his plunge into origin of life research "the second biggest thrill" of his life. Sadly, the field has lost his boundless enthusiasm. Harry Lonsdale died just weeks ago at age 82.

Lonsdale was a retired chemist and wealthy businessman who ran three times unsuccessfully for the US Senate from Oregon. He was looking for creative chemistry in the 76 proposals on origin of life he received between 2011 and 2012 in response to his announced competition.

In June 2012, Lonsdale began awarding research money to his prize winners -- British chemists John Sutherland and Matthew Powner. An American team and Canadian team were also recipients of Lonsdale research funds. The three teams were refunded in 2013, and again in 2014.

Harry Lonsdale did not publicly share information about the proposals that failed to make it to "the top of the heap," papers that could be worth further scrutiny. Our interview, "The Harry Lonsdale Prize," is the book's opening chapter.

But while Lonsdale's initial intention may have been to figure out the origin of life, it has become increasingly clear that we can't go back in time to that precise moment of life's emergence because so far we don't have the tools to determine what the exact conditions were (assuming we live in a 3D world, Fermi Lab is now checking if it's 2D -- a hologram!).

What Freeman Dyson, Institute for Advanced Study emeritus professor (Chapter 8), told Charlie Rose over the PBS roundtable several years ago still holds: "we are all equally ignorant" when it comes to origin of life. Or as synthetic biologist Steve Benner (Chapters 4, 6) puts it, "there are no experts." This has been the state of affairs ever since the seminal work of Soviet biochemist Alexander Oparin in the 1920s with his conceptualization of droplets of life he called "coacervates."

As a result, my interest in the field continues to be the politics of the investigation into origin of life as much as the science itself.

Indeed, it is because of the sense of futility in understanding the exact origin of life that some scientists decided to move the goal to that of making life in the lab, *i.e.*, synthetic cell development, with the rationale that making a "toy model" of life could be the best way to piece together the puzzle. They call their research "origins of life" rather than "origin of life" to reflect that there may be an indefinite number of possibilities.

The three bottom-up approaches to making a synthetic cell are: (1) a protocell from scratch (Szostak, Sutherland *et al.*), (2) a minimal cell built from existing natural parts consisting of 200-400 genes (Vincent Noireaux's lab at the University of Minnesota, *e.g.*), (3) an artificial cell made from artificial parts (Steen Rasmussen at the University of Southern Denmark *et al.*).

The push to make a protocell is only a decade or so old. For example, in 2002, four scientists -- physicists Norman Packard and Steen Rasmussen, chemist John McCaskill and philosopher Mark Bedau (dubbed the "Four Protocell Musketeers" by author Ed Regis) made a pact to build a synthetic cell with $14M in research funds from Los Alamos, the European Commission and Packard's own pockets. In my interview with Rasmussen (Chapter 13), we discuss how his estimate that we'd have a protocell within a decade was obviously wrong. Rasmussen believes it is now "very close" to happening.

Norm Packard (Chapter 13) tells me the money for his ProtoLife enterprise eventually dried up in Europe, so he moved the company to the US in 2008. But America's market turndown at the time led to a reorientation of the company. As I write, ProtoLife is no longer in development on a synthetic cell and some of Packard's former researchers have moved on to other synthetic cell labs.

The anointed scientist of the origin of life field is currently Harvard University biologist and Nobel laureate Jack Szostak. Astrophysicist Mario Livio goes further, dubbing Szostak, "The Leader of Origin of Life Studies in the Universe."

Szostak sees genetic material, *i.e.*, RNA (ribonucleic acid) that can replicate as essential to his protocell. He is trusted and has the respect of his peers because his science is formidable -- although some in rival camps regard him and his allies as "the enemy."

Another revered figure in the origin of life community is biochemist Pier Luigi Luisi. Following the official meeting of ISSOL (International Society for the Study of the Origin of Life and Astrobiology Society) this year in Nara, Japan, Luisi chaired the unofficial meeting on "open questions."

Luisi was relatively unmoved by the ISSOL pronouncements and now thinks that we need to "come away from the DNA or RNA-centered view of life." That we need a fresh start, all new "mindstorms."

Says Luisi:

> "[T]he prerequisite for the origin of life is the onset of kinetic control -- namely catalysis -- something which permits a departure from thermodynamic control." (Part 6 -- "Rethinking the Circus")

Nevertheless, in May 2014 at the World Science Festival in New York, Jack Szostak announced he hoped to have "life in the lab" -- or "Generation II," as Harvard astrophysicist Dimitar Sasselov calls it -- in three to five years (closer to three years). Szostak confirmed to me that he's working bottom up, from scratch and that John Sutherland, Matt Powner and others are being consulted on the ingredients he will put into his lipid (*i.e.*, fatty) vesicles in the lab.

Dimitar Sasselov told me that part of his contribution to the life in lab experiment is providing the right ultraviolet radiation. He's now building a small lab to simulate conditions of early Earth or of some of the exoplanets.

My conversations with Szostak, Powner and Sasselov are featured in Part 4: "How to Make a Protocell." John Sutherland has ducked requests to be interviewed and has also asked that I not quote from his 2013 video presentation linked on Harry Lonsdale's web page (www.originlife.org). It's an hour of Sutherland on Skype, the master chemist revealing his origin of life brew in the making.

I recommend watching this not-ready-for-prime-time show, including the session break featuring Georgia Tech chemist Nicholas Hud as one of Harry Lonsdale's project reviewers. Nick Hud's own origin of life research is funded by NASA and NSF. Hud thinks you don't get RNA abiotically, that other molecules came first ("Nick Hud, the Lion Tamer," Chapter 4).

The video presents two other Lonsdale-funded teams as well, that of ISSOL president Dave Deamer *et al.* and Portland State University's Niles Lehman and his colleagues.

Curiously, Deamer tells me (Part 1: "Who Owns Orign of Life?"), that Lonsdale told him Sutherland thinks it will take 50 years to get to the bottom of the origin of life. Sutherland's colleague, Matt Powner estimates that it will take 10 - 20 years to make the "missing link" between prebiotics and life as we know it (Chapter 11) . . .

Whether you call it "origin" or "origins of life," the field now has an array of visionaries. "Origins," in particular -- figuring out the science of making toy models, *i.e.*, synthetic cell development -- is what is attracting private money. But only recently.

Dimitar Sasselov, who directs the Harvard Origins of Life Initiative, reveals in our interview his struggle to secure origins funding early on. Says Sasselov:

> "We told the [Harvard] provost committee [in 2006], 'We need you to share our vision and support our research because the federal agencies -- NASA, NSF, NIH --

won't.' We added that we couldn't predict whether it would work out."

Now playing a central role in design of the protocell through its philanthropy is the Simons Foundation, a private organization incorporated in 1994 and headquartered in Manhattan, although the foundation does not publicize its support as such. In fact, James Simons, chairman of the foundation and also chairman of Renaissance Technologies -- the hedge fund that enables his philanthropy -- told me in April at a Simons-sponsored event that he knew nothing about origin of life when he invited me to stop by the foundation's Fifth Avenue offices for an interview and tour.

I was delighted that Simons later decided to share his thoughts with me on the subject following his summer at sea aboard Archimedes, his 219-foot toy boat (Chapter 11).

The Simons Collaboration on the Origins of Life (SCOL) has pledged eight years of financial support to teams of investigators and postdocs, as well as to its steering committee and directors.

Each SCOL investigator receives annually between $100K and $400K plus 20% for expenses. Each SCOL postdoc receives $80K annually (inclusive of health care coverage and other expenses). Awards to the SCOL steering committee and co-directors -- Jack Szostak and Dimitar Sasselov -- are more. How much more is not apparent. All origins research is done outside the physical space of the Simons Foundation.

Jim Simons, and his wife Marilyn -- who started the foundation and serves as its president -- have seen to it that women share substantially in their charitable contribution to basic science. SCOL investigator Lisa Kaltenegger has just received a Simons award of $1M, *e.g.*

Jim and Marilyn Simons were among the first to sign the Giving Pledge, meaning they plan to give away most of their wealth in their lifetime ($12.5B).

Meanwhile, DARPA (Defense Advanced Research Projects Agency), whose motto is "creating and preventing strategic surprises," has also gotten into the act, by way of minimal cell development -- involving *E. coli*. DARPA is funding Vincent Noireaux's new lab at the University of Minnesota, which I visited last spring (Chapter 12). Hopefully, the agency is looking for a way to deal with crises like drug-resistant superbugs, etc., rather than creating surprises.

Other visible financiers of the origin of life field include the John Templeton Foundation and The Jeffrey Epstein Foundation.

Germany's Max Planck Society had been actively seeking a director and staff in 2014 for a separate origin of life institute but recently put the project on hold after failing to recruit the chemist it thought could lead the breakthrough research. It is unclear, as I write, who that scientist was and whether they declined the Max Planck Society offer or if the Society found the scientist's strategy lacking.

Japan has ELSI, the Earth-Life Science Institute looking for origin of life everywhere from the core of the Earth--beyond, as my interview with ELSI director Kei Hirose reveals (Chapter 7).

Even Hollywood is putting money into exploring for clues to life's origin with, for instance, the recent James Cameron 3D movie about a submarine dive to the deepest point on the planet . . .

About halfway through my research, in April 2013, I decided to get away from it all and New York and just let life wash over me. The move followed a year's avalanche of academic papers and growing stacks of audio tapes, audio tape transcripts and yellow pads (the latter seemed to be self-replicating) from phone interviews with scientists in various parts of the world and travel to origin of life conferences.

I first flew to the American Southeast, and then to the Northeast, Northwest, Southwest and other destinations. I lived for two months on an island in South Carolina's lowcountry, part of a protected marshland, a pristine salt system where the daily drama is one of legions of minute fiddler crabs charging up and down the oyster-encrusted river banks, with an occasional egret or ibis walking on by. Where families of dolphins thrive and so do hawks and osprey.

The beaches, some ghost-forested and adjacent to a former cotton plantation, are patrolled by squadrons of pelicans flying over a world of sea grass, sponges, whelk egg casings -- some chains three-feet long and bearing a shocking resemblance to the human vertebral column -- sea cucumbers, jellyfish and myriad other invertebrates resting for a while in the foam and froth or maybe forever. Beaches where perfectly whipped clouds appear at sea level and take turns lighting up a stormy night. Where sleeping on the screened-in porch feels like the *rififi* of first life.

Images from those months remain with me more so even than the display of life I saw in subsequent weeks living on the lobster island of Vinalhaven, which the American bald eagle still calls home, or in the San Juan archipelago of the Pacific Northwest where extraterrestrial-sized starfish and bull kelp (tipped with floating orbs of carbon monoxide gas, etc.) are reminders that we are not alone.

Science, at 400 years old or so, cannot really articulate the marvel of such a montage of life billions of years in the making, or yet adequately explain how it came to be. But I decided that the efforts now underway by scientists to sort it all out, to take on the challenge of understanding the origin of life was not just a great story to tell. It was the greatest.

So, following my last outpost on Wasp Pass, seven miles from the Canadian border, watching seals splash around outside my kitchen window, and another few weeks at scientific conferences in Europe with a brief return to the San Juan Islands and a stop at the Santa Fe Institute -- I came home to New York to finish my book.

Home, down the street from John Belushi's old haunt, poet laureate Joseph Brodsky's place and around the corner from Edna St. Vincent Millay's narrowest house in New York on Cherry Lane -- where there's a red plaque bearing her famous words: "[M]y candle burns at both ends" (but none across the street at the former CIA safehouse, site of the infamous MKULTRA experiments).

It took months to process what I'd been through traveling . . .

In addition to Harry Lonsdale, two other scientists who are an important part of this book have died -- Lynn Margulis and Carl Woese. Lynn Margulis is best known for her popularization of endosymbiosis theory and eukarya, the organism that gave rise to the other multicellular organisms. Margulis received the US Presidential Medal for her achievements in science in 1999. I include my story about Margulis's challenge to the *Proceedings of the National Academy of Sciences* over its secretive peer review policy (Part 7, "It's the People's Circus").

Biophysicist Carl Woese many consider the most important evolution scientist of the last century, and Woese was, indeed, honored with the major prizes in his lifetime -- except for the Nobel, perhaps because he was truly a revolutionary. In our conversation (Chapter 15), for example, which may be the last feature interview he gave, Woese lamented not having "overthrown the hegemony of the culture of Darwin."

Carl Woese is perhaps best remembered for his identification of a "Third Kingdom of Life," the archaea -- a methane-producing microbe that lives in a range of habitats, including the gut of modern animals. Like other microbes, the most prevalent form of life on the planet, archaea rely on the lateral transfer of information.

University of Chicago microbiologist James Shapiro and Oxford's Denis Noble, who now have a web site called "The Third Way of Evolution" (www.thethirdwayofevolution.com), pick up where Woese left off on paradigm shift in my separate interviews with them (Chapter 15).

Woese also told me he considered the collaboration with colleague Nigel Goldenfeld at the University of Illinois, Urbana-Champaign the most productive one of his scientific career. Goldenfeld, a condensed matter physicist, and Woese, along with their team at UI-UC's Institute for Universal Biology (IUB), were awarded $8M by NAI to research the principles of the origin and evolution of life just three months before Woese died. Goldenfeld now heads IUB. In our interview (Chapter 15) Goldenfeld makes the case for a consensus on what life is, for a general theory of life.

I had almost enough research for a book when I returned to New York from my odyssey, but still needed to talk to the "origins" scientists who are actively making life in the lab, Szostak *et al*.

Scientists tend to shy away from being definitive about just why they're attempting to synthesize life, except to say that it's an interesting intellectual exercise. This, of course, leaves the door open for what society will then do with this synthetic cell once scientists make it. And though scientists may call it a toy model, once that toy model is a done deal, it is no longer a toy. As Dimitar Sasselov reminds in *The Life of Super-Earths*, it is "a life-form that has no place on Earth's tree of life, a new life-form at the root of a new tree of life."

Yet there is still no agreement even about what existing life is, as mentioned above. The emerging thinking is that life is not gene-centered -- even as scientists cling to the comfort of a code -- life is more relational, systemic.

Jack Szostak says he does not think defining life is so urgent when it comes to making it in the lab. He comments in "Attempts to Define Life Do Not Help to Understand the Origin of Life" (*Journal of Biomolecular Structure & Dynamics*):

> "An inordinate amount of effort has been spent over the decades in futile attempts to define "life" -- often and indeed usually biased by the research focus of the person doing the defining."

When I asked Jack Szostak what kind of life he was making in the lab, he told me this:

> "I don't worry, at this point, whether we'll know exactly how it happened on the early Earth. What we're trying to do is work out a plausible pathway where all of the steps seem chemically and physically reasonable, and maybe we'll end up with multiple pathways which are all possible."

"[M]ultiple pathways . . . all possible."

Meanwhile, we have an ongoing parade of origin of life acts -- hypotheses, often presented in impenetrable technical language, that keep out wide public scrutiny. Papers some scientists have a hard time reading. Papers for which there are no solid benchmarks for what is plausible, borderline creation myth, hallucination or charlatan seduction.

Sometimes there is no common ground at all between origin of life scientists. Crucial UV light for one scientist's model can mean death for another's, for example.

What's more, the disagreements can be fierce, despite what Neil deGrasse Tyson told *Parade* magazine about scientists being a peaceful lot. And that's just what we see in public. The private lives of publicly-funded scientists have not been of much media interest.

One clash I report focuses on the compartmentalist camp view of how life could have emerged with the right energetics, the right "oomph," inside metal membranes of ancient alkaline springs (if they ever existed) similar to today's Lost City vent system. Lost City is so far a unique structure, sitting atop the Atlantis Massif in the Atlantic Ocean. Michael Russell, Elbert Branscomb, Nick Lane, Bill Martin all weigh in (Chapter 9).

There are scientists who claim origin of life is 100% physics and others who insist it is all chemistry. Inorganic chemists trying to make matter come alive who scoff at the bravado of organic chemists attempting to cook-up life, while organic chemists deride the artificial life effort as "false flag". . .

Also sharing the Big Tent is the top-down approach to origin of life, *i.e.,* robotics, which seems to have more support in Europe and Japan than in the US -- aside from its fans at Singularity University, who think people will not be able to survive in the future if they don't embrace robotics. I talk to Jaron Lanier, "the father of virtual reality," about this (Part 5, "Top-Down Tent"), and also Oxford philosopher Vincent Müller, as well as cyberneticist Slawek Nasuto, co-creator of the "animat" (a closed loop -- part animal, part robot).

11

One of the more spectacular Big Tent acts (if it exists) is "Poised Realm." Investigators Gábor Vattay, Stuart Kauffman, and Samuli Niranen think they've found a state of matter that hovers between the quantum and classical worlds that could be useful in "drug discovery, computers, and artificial intelligence." Stu Kauffman says more, much more, in our interview (Chapter 5, "Waiting for Ganesh").

Hear the calliope? Come ride the circus carousel. . . . Clowns? Of course there are clowns.

Suzan Mazur
November 2014

Part 1

Who Owns The Circus?

"First of all -- what is life? My own home-style definition is this: (1) something that can reproduce itself, and (2) live off the land. It must be able to extract useful energy from its environment. We do that by eating mashed potatoes. But first life did not do that." -- **Harry Lonsdale**, Origin of Life Challenge philanthropist

Chapter 1

The Harry Lonsdale Prize

HARRY LONSDALE
(photo, courtesy H. Lonsdale)

Remembering Harry Lonsdale, who died November 11, 2014.

As a long-time protector of the Pacific Northwest's old growth forests and the once political darling of environmentalists there, philanthropist Harry Lonsdale in recent years thought there was no greater wilderness to get his feet wet in than the investigation into origin of life.

Lonsdale was a chemist who became a millionaire following the sale of his Oregon high tech company to Pfizer in the 1980s and then ran three times (unsuccessfully) for the US Senate. In 2011, he made an international splash announcing in leading science magazines an Origin of Life $50K Challenge he was funding, reaching out for the best proposal detailing "first life" (www.originlife.org).

In June 2012, Lonsdale announced the $50K winner and awarded an additional $300K in research funds to be split among the top three finishing teams. He said he was looking to give away a total of $2M in support of origin of life investigations in the months to come in increments of $300K.

In his search for a winning hypothesis, Harry Lonsdale focused on proposals he thought cogently addressed the chemistry of first life, an area of research where he said he and his panel of peer reviewers were most knowledgeable.

Lonsdale told me that -- per the industry practice -- he'd been "sworn to secrecy" not to disclose the names of the reviewers, "the Pros," as he called them, later remembering that he'd already revealed three of the names to an online magazine.

Not everyone was satisfied with Lonsdale's embrace of anonymous peer review or with the angling of the Challenge to Darwinian science.

Still, Lonsdale went where few but NASA, and now the Simons Foundation, have gone in funding origin of life research. And Lonsdale was again applauded as a champion.

Origin of life investigation is one of those areas of research where amateurs and credentialed scientists are on somewhat equal footing, since the terrain is still so fresh. So I was curious as to why no amateurs were among Lonsdale's top finishers, and what private money thrown into the mix might do that public money had not been able to. I also wondered whether Lonsdale's political savvy was enough to deal with the politics of origin of life science.

Harry Lonsdale's PhD was in chemistry from Penn State. He was the author of the book *Running: Politics, Power and the Press* and of more than 100 scientific papers and various patents, as well as founding editor of the *Journal of Membrane Science*.

Lonsdale's was a rags-to-riches story. He was brought up on a chicken farm in New Jersey. His mother was a Sicilian immigrant and his Welsh father orphaned on the doorstep at age two.

I had several phone conversations with Harry Lonsdale over the last two years. His enthusiasm for origin of life was boundless.

May 2012 conversation

Suzan Mazur: Your initiatives regarding campaign finance reform as well as the environment in the Pacific Northwest, your run for the US Senate from Oregon and your anti-war activism have been greatly appreciated. Jeffrey St. Clair, the environmental journalist and editor of *CounterPunch*, asked me to convey that you still have a lot of fans in "Orygun."

Harry Lonsdale: That's kind of him to say so. I know Jeffrey's writing. He's a great guy.

Suzan Mazur: Now that you're living in California, I'm curious if you might be thinking about another run for public office sometime soon.

Harry Lonsdale: HEAVENS NO. You can quote me on that for sure. Heavens no. Let me just clarify. It's the hardest thing I've ever done by far. The days are long. The pressure is endless. The rejection sometimes is overwhelming. So the answer is no.

I made many friends campaigning, for which I'm grateful. I hope some lifetime friends, scattered pretty much across Oregon, some around the country. But it was very, very hard on me. I'm not a particularly backslapping gregarious type.

Suzan Mazur: You sound gregarious.

Harry Lonsdale: I like people but I'm not a crowd person. I tend to be a wallflower.

Suzan Mazur: Why did you leave Oregon for California?

Harry Lonsdale: I spent the last 14 or 15 winters in the south, in Arizona and here in California. I got tired of Oregon winters. I lived in Bend, Oregon, which has pretty long winters, not severe, but it snows a fair amount. It's colder than I like. I'm going back to Oregon for the summer. I still love Oregon.

Suzan Mazur: And now you're championing an investigation into the origin of life. . .

Many think prebiotic evolution is too speculative to be discussed seriously, including University of Chicago microbiologist James Shapiro, who I recently interviewed. Others like the 1,000 or so researchers in the NASA astrobiology program no doubt disagree. By your private funding of the Origin of Life Challenge, what do you hope to discover regarding origin of life that public funding has not been able to?

I'd like to note here your definition of life, which is, "a self-sustaining chemical system capable of undergoing Darwinian evolution."

Harry Lonsdale: Let me just wade in. First of all -- what is life? My own home-style definition is this: (1) something that can reproduce itself, and (2) live off the land. It must be able to extract useful energy from its environment. We do that by eating mashed potatoes. But first life did not do that. First life was somehow able to extract useful energy from a very dismal environment three billion plus years ago. We don't quite know how yet.

Suzan Mazur: But whose definition is the one I just quoted, which is on your Origin of Life website?

Harry Lonsdale: Not mine. That definition for life is the standard definition among the experts. And those five or six questions are not mine:

> ➢ "What were the nature and genesis of the first macromolecules on the prebiotic earth?

- How did the building blocks that comprise these macromolecules become available and how were they assembled?

- How did prebiotic molecules first acquire the capacity for storing genetic information and how did the genetic machinery evolve?

- At what stage in the origin of life did cells originate, and what did they contain?

- How did those primitive cells evolve to modern biological cells?

- What was the chemistry of the first metabolic pathway(s) and how did that metabolism evolve to modern cellular metabolism?

- At what stage did proteins become involved in metabolic processes and how did the link first arise between genetic molecules and other functional molecules, such as enzymes?"

I got those from some of the Pros, particularly Jerry Joyce at Scripps Research in La Jolla, California. He was the principal author of those words. He and Dave Deamer at UC - Santa Cruz helped me with those five or six quotes. They're not my words at all.

Suzan Mazur: The National Academy of Sciences has urged the US government to endow incentive prizes of tens of millions of dollars. Do you agree that more philanthropists should step forward to fund origin of life investigations even though how life began could for some time remain a "guess," as David Deamer, who heads ISSOL (International Society for the Study of the Origin of Life and Astrobiology), characterized his own view of the subject in a conversation with me.

Harry Lonsdale: Yes, I do agree that more philanthropists should step forward to fund origin of life investigations. I'll elaborate, but let me first back up. You asked me why I'm doing this in view of the fact that there's already some NASA support for it. The people I've talked with, in fact the people I've made grants to, all say money in this field is extremely hard to come by.

You might think the National Science Foundation would have a great interest in this. But they don't. Their funding of origin of life research is minimal. They're funding a group at Georgia Tech -- a man named Nicholas Hud has funding there -- but beyond that I think there's almost nothing.

Origin of life is one of the greatest unknowns we still have in science. Five hundred years ago most of how the world and Universe worked was a mystery. Now many of those mysteries have been solved. Three of the great remaining mysteries are: (1) origin of the universe, which we may never solve; (2) origin of life, which is a huge problem in my opinion; and (3) how the brain works.

Origin of life is largely underfunded. And there's almost no international funding. One of my prizewinners is a Canadian-US team -- which I'm not going to name right now -- and they told me there's no money for cross-border research. That is sad. There's a lot of money in this country, there are a hundred or more billionaires.

Bill Gates and Warren Buffet put together "The Giving Pledge" three or four years ago. They signed up about 70 people, all of them centi-millionaires and up, who have agreed to give away most of their wealth for worthwhile causes. Most of the money is going to hospital extensions and scholarship for students. Great stuff -- but that's not unveiling any of the great remaining mysteries. Some of those funds should be channeled to scientific endeavors.

Ninety percent of the dough for origin of life research in the US comes from NASA. As a result, these researchers funded through NASA bring to their work the idea that life, or pieces of life, came from outer space. And it may have. But I don't think there's any evidence that pieces of life could not have been produced here on Earth. Yet there's a bias in US origin of life research toward the

19

extraterrestrial -- space stuff, meteorites. I'm not sure such an emphasis is essential to discovering how life began on planet Earth. It's certainly not part of my thinking.

Suzan Mazur: Bruce Runnegar, the former NASA Astrobiology Institute chief, who's now a professor of paleontology at UCLA, seems to agree with you. He told me regarding origin of life research that "investigations and experiments are relatively inexpensive on Earth. You can do all this for a few tens of thousands of dollars. But it's a highly different matter if you're going to Mars and spending a billion or two billion dollars returning samples or doing experiments *in situ*."

Harry Lonsdale: One more point to make. I'm holding in my hand right now a full page ad from the *New York Times* by Bonhams international auctioneers and appraisers. It's a photograph of a beautiful vase. Bonhams auctioned this vase off for $14,332,000. That's a ton of money. Now I don't know where that vase is going. I presume it's going to be sitting on the shelf in some museum or somebody's livingroom. But for $14M we could fund all the origin of life research in the United States for five or 10 years and we'd still have the vase. A much better use for that money is scientific research, whether it's origin of life or otherwise. It's my lifelong training as a scientist to feel that way, I suppose, but we have money in this country and we should use it for some useful purposes.

Suzan Mazur: Are you currently designing any other prize, science or otherwise?

Harry Lonsdale: No.

Suzan Mazur: There has been criticism of science prizes. It's been suggested that the prize money is often a platform for the sponsor of the prize. That a lot of the genius winners may be no more deserving than the next intellectual and that the prizes are frequently awarded to those whose ideas advance a certain belief system, *i.e.*, the process is corrupt. Would you comment?

Harry Lonsdale: I would. In some cases that may be true. It may be a platform for other people, I don't think it's a platform for what I've done. I announced my prize, I bought space in *Nature*, *Science* magazine and *Chemical & Engineering News* to advertise the Challenge. And I printed a thousand brochures for distribution around the world. I attracted 76 proposals from 19 countries. I also assembled a team of peer reviewers.

Suzan Mazur: You first approached Dave Deamer in 2010 at UC - Santa Cruz with the proposal for the Origin of Life prize, and Deamer thought it was a great idea considering government funding for origin of life research has been drying up. Can you tell me who, aside from Dave Deamer, NASA - Ames planetary scientist Chris McKay and Nobelist Jack Szostak were referees for the prize? Were there any women on the peer review panel?

Harry Lonsdale: I have to say no, Suzan.

Suzan Mazur: Was it racially and ethnically diverse?

Harry Lonsdale: I'm afraid it was all white males.

Suzan Mazur: And was it a panel of international experts?

Harry Lonsdale: Yes. One from Europe and seven from the US.

Suzan Mazur: Can you tell me who they were?

Harry Lonsdale: I'm sworn to secrecy on that.

Suzan Mazur: Is that right?

Harry Lonsdale: Those [three] names that you have, by the way, are correct -- but I don't want to reveal the other names because. . . NSF proposals are reviewed by peers and their names are never disclosed, so I'd rather not. But trust me, they were good people. I asked the best people I knew in the field who were the experts. I asked three such people, they all gave me a list. I combined those lists and picked out the 10 best people I could think of. I called 10. Eight accepted. So I immediately had a peer review panel.

Suzan Mazur: Were the referees paid?

Harry Lonsdale: No. I picked up their expenses to come to a meeting to discuss the proposals, but that's all. I think referees are almost never paid as far as I know. They were eager to help all the way along the line. A group of great integrity.

Suzan Mazur: How did the selection process work? Did the referees select the winners or did you? And when will we know who won?

Harry Lonsdale: The panel reviewed the top dozen proposals and then made recommendations on those proposals to me in writing. But in the end, it was my call.

On March 30 of this year I met from 9 in the morning until 4 in the afternoon with all the reviewers at a hotel in San Diego. They came from across the US and from Europe. We discussed the top proposals and the reviews they had written beforehand. We all sort of sat around and chewed the fat. I asked a thousand questions. I then chose my favorite three proposals.

The three I picked were three of the four the experts had recommended. So I largely leaned on the reviewers, but ultimately the decisions were mine.

There are three groups of winners. They're all teams. No individuals. The $50K goes to a team of two and an additional $300K in research funds will be shared by scientists from all three groups.

Suzan Mazur: When are you going to announce?

Harry Lonsdale: Before the end of this month we'll put out a press release. I'm working on the press release right now. But until that's done and sent out, I don't want to discuss the winners.

Suzan Mazur: All teams. A dozen people or so?

Harry Lonsdale: It's actually seven people on three different teams and seven institutions. Seven people, seven institutions, three proposals

Suzan Mazur: All institutional researchers.

Harry Lonsdale: Yes.

Suzan Mazur: No independent scientists.

Harry Lonsdale: No. A lot of independent people proposed but somehow their proposals didn't get to the top of the heap.

Suzan Mazur: Any hints about the content of the winning proposals?

Harry Lonsdale: The papers are all based on the RNA world. How first life, how this first creature lived off the land is not addressed, only how that individual creature came to be and how it was able to reproduce itself. The thinking now is that RNA was the precursor of DNA. It's a much simpler molecule.

All three winners look at some aspect of how RNA came to be and how it evolved. RNA has catalytic activity like proteins have. But in early life it's thought, again not by me but by the experts, that proteins didn't exist. So what was the catalyst? The thinking is that the catalyst was an RNA molecule called a ribozyme, which is a long strand of RNA. All the work I'm supporting deals with this RNA world -- where RNA came from and how it came to be an early catalyst.

Suzan Mazur: The guidelines for the prize were as follows: "The proposal should take into account the conditions, materials, and energy sources believed to have existed on the prebiotic Earth. Submission should provide a cogent hypothesis for how life first arose, including its plausible chemistry, and for how primitive life could have evolved to modern biological cells, including the present genetic material and metabolism."

Your PhD is in chemistry and your undergraduate degree also.

Harry Lonsdale: That's correct.

Suzan Mazur: How deeply involved in origin of life science are you? And in designing guidelines for the prize, do you think that you and your advisors underplayed the importance of physical processes in prebiotic evolution? I'm thinking about, say the work of Duke University mechanical engineer and Constructal theorist Adrian Bejan and his investigation of river basins and prebiotic flow systems, etc., where the line between animate and inanimate is blurred and life is viewed as an organized flow of matter, electricity, heat, etc.

Harry Lonsdale: Can I back up before I answer that question. That sentence you read to me over the phone as to what I hoped it was people would give me in terms of their proposals, I must say I was very much underwhelmed by the breadth of their proposals. **Even the experts I drew together in San Diego a month or so ago, even they don't have a single clear model of how life began.** There's no universal agreement. We don't have a theory. We're a long way from home base. Probably 10 or 20 years away before we have a plausible model and even further out into the future before we can say we know how life began. I'll be dead before people can make that statement.

Suzan Mazur: Is it because your background is in chemistry that you were more interested in proposals based on chemical approaches to origin of life?

Harry Lonsdale: I would say that of the eight people on my panel, three at most were trained in chemistry, some were biochemists, there was one physicist, a space scientist and two biologists -- it was a pretty broad range of expertise. No engineers. I guess we underplayed the physical processes, but not intentionally.

Suzan Mazur: But did you get proposals along those lines?

Harry Lonsdale: A few. We had proposals that came in from every direction.

Suzan Mazur: Chris McKay once told me the following:

> "The Darwinian paradigm breaks down in two obvious ways. First, and most clear, Darwinian selection cannot be responsible for the origin of life. Secondly, there is some thought that Darwinian selection cannot fully explain the rise of complexity at the molecular level. . . . It can't be Darwinian all the way down. . . . Darwinian

selection only works when there's software. And everything that's prebiotic is hardware."

Again, "life" has been defined on the Origin of Life Challenge website as "a self-sustaining chemical system capable of undergoing Darwinian evolution." My question is that by steering your prizewinning search for the origin of life in the direction of Darwinian science, which is now being marginalized in light of the "evo-devo revolution" -- as Noam Chomsky calls it -- and an evolution paradigm shift, with some of our most esteemed scientists declaring neo-Darwinism dead -- the accumulation of genetic mutations being enough to change one species to another having not been validated in the literature -- are you concerned that you and your panel may have angled the prize in the direction of false hypotheses?

Harry Lonsdale: I think that first life was capable of evolution and that evolution began on that day that first life came into being. Was there evolution before that first life? I don't think that evolution is what brought that first life to you and me . . .

We were hoping people might submit proposals covering the gamut, from first life to modern life. No proposals attempted to do that. It's too big a question right now. It's 2012 -- let's wait until 2025. Maybe by then people will have put the whole puzzle together, but right now we're looking at pieces, or pieces of pieces, of this puzzle.

Suzan Mazur: What I'm questioning is angling of the prize to Darwinian science.

Harry Lonsdale: When you say angling toward, it's true. First life was angling toward but not yet there. Evolution came after first life.

Suzan Mazur: I asked Dave Deamer if life had a beginning or is it just part of a process inherent to the universe. And he said, "It's part of a process." I also asked him if evolution started when the universe was born. His response was "It depends on what you want to call evolution."

Harry Lonsdale: Evolution is a process once life exists, is how I would put it.

Suzan Mazur: Life has been on earth for nearly four billion years, how urgent would you say it is that we now find the answer to the great mystery of who we are and where we came from?

Harry Lonsdale: Why now? That's a tough question to answer. Life on Earth is almost four billion years old, most of what we know about the rest of the universe is about 400 years old. It really started with Isaac Newton and Copernicus. An enormous amount has happened in the last 50 plus years since the discovery of the structure of DNA, and the rate of increase in knowledge is accelerating all the time with the Internet.

I suspect we're going to know more in the next 50 years than we knew in the previous 500. So as to the urgency of discovering who we are and where we came from -- I'd say what's important to focus on is not that life has been unaware of its origin for nearly four billion years and we've gotten along just fine, but to consider the possibilities once we move beyond these hard questions and finally make the breakthrough.

I've had four careers in the last 80 years. Starting my own company was the biggest thrill. But this is the second biggest thrill of my life. It really is.

Chapter 2

The Pack & The Prize: Lawrence Krauss

LAWRENCE KRAUSS
(*artist, Peter Sheesley*)

"Loaded with bias is the review process reserved for the big projects. The review is run by the "leaders," the persons who head (or have headed) the big projects. They are the influential, the ones who are consulted during the review process and even before a new research initiative is selected for funding. . . They are many, not one. They constitute a social stratum known colloquially as academic mafias and dark networks (in social dynamics, these terms mean "networks of persons exerting hidden influence"). Favored are the applicants who work for the mafia." -- **Adrian Bejan**, *International Journal of Design & Nature and Ecodynamics*

June 19, 2012

We like Harry Lonsdale and his generosity to science. But Lonsdale's Origin of Life Challenge dollars, $200K of them, went to a team of British academics, a red flag perhaps that American researchers increasingly prefer the stage to the lab, as well as evidence of the hand of "academic mafias" at work in the prize process.

However marvelous the experiments of John Sutherland, half of the winning team of John Sutherland-Matthew Powner, he and his RNA work were showcased three years ago in a series of stories by Nicholas Wade in the *New York Times*. Sutherland was then featured in a PBS *Nova* program along with a couple of Lonsdale's peer reviewers (one of whom received research dollars from Lonsdale as part of the Challenge package). Then Sutherland received an award for origin of life research from the Royal Society, now headed by Paul Nurse, formerly president of Rockefeller University and host of the May 2008 symposium: "From RNA to Humans" featuring two of Lonsdale's peer reviewers.

What is certain is that the 70 losers of the Lonsdale prize stood little chance of making it to the top of Lonsdale's heap of origin of life proposals.

Harry Lonsdale says he has no plans to publish the non-winning proposals or to release the names of their authors. So I went to the gatekeeper of the Lonsdale prize, Arizona State University's Origins Project (ASU-OP) director Lawrence Krauss, an astrophysicist, for further clarification as to how things were organized.

Krauss is a public intellectual who currently amuses audiences by pulling a universe from out of nothing -- at least in theory. But then Krauss has amassed a body of work impressive enough for at least two distinguished scientists, a visionary who clearly has something to say about nothing.

I spoke with Lawrence Krauss by phone in Australia where he's been promoting his latest book, *A Universe From Nothing*, as well as the virtues of atheism. He was next moving on to the UK to do a film with Richard Dawkins, then swinging through New York to appear on *Colbert*.

Aside from his role at ASU-OP, Lawrence Maxwell Krauss is Foundation Professor of the School of Earth and Space Exploration and serves as associate director of Paul Davies' Beyond Center at ASU.

Last year Krauss and ASU-OP agreed to partner with Harry Lonsdale to assist with the awarding of Lonsdale's pledged $2M for origin of life research and to raise public awareness about origin of life. ASU-OP grew out of an origins symposium Krauss organized in 2009 attended by 3,000 people, including 80 scientists.

Lawrence Krauss's PhD is from MIT in physics.

He is the author of 300 studies, and these eight books: *The Fifth Essence*; *Fear of Physics*; *The Physics of Star Trek*; *Beyond Star Trek*; *Quintessence Atom*; *Hiding in the Mirror*; *Quantum Man*; *A Universe From Nothing*.

My interview with Lawrence Krauss follows.

Suzan Mazur: When did your partnering begin with Harry Lonsdale's Origin of Life Challenge? Lonsdale did not mention you during my recent interview with him, and I didn't see any reference to you and/or the Arizona State University Origins Project (ASU-OP) you head in the original Lonsdale outreach for proposals.

Lawrence Krauss: The Lonsdale proposal happened before the partnering. I became aware of Harry or vice versa after the proposal was announced. I forget who made the first phone call to discuss the fact that we had established a vibrant program the Lonsdale prize would mesh perfectly with. But either way it became clear to both of us that it was a natural fit.

Suzan Mazur: When did the partnering begin?

Lawrence Krauss: It was probably a year ago.

Suzan Mazur: What does the partnership entail?

Lawrence Krauss: The partnership began with Harry getting involved on the advisory board of ASU-OP so he could learn more about the ongoing program. He came to ASU three or four times in the last year, meeting with me and the president of the university and other individuals to discuss ways in which we could facilitate the giving of the prize and administration of the money afterwards. We agreed to assist with announcement of the prize and to encourage others to support origins of life research. So the partnering entails using resources of ASU-OP and the university to raise public awareness about origins of life research and generate interest in the philanthropic community as well, to match and add to the support that Harry's giving.

Suzan Mazur: Were you involved in the selection of the peer review panel?

Lawrence Krauss: No. We consulted him to let him know how good the group was, but Harry did that all himself.

Suzan Mazur: Now that the Lonsdale prize has been awarded, can you say who the peer review panel members were, aside from Dave Deamer, Chris McKay and Jack Szostak, and whether the same referees will be advising regarding future origin of life Lonsdale grants? Lonsdale said he plans to award an additional $1.65M.

Lawrence Krauss: The peer review panel is a question for Harry to answer. The plan is for the money to go out over a six to seven-year period. It's contingent on work being done and progress. The exact framework has not been finalized. The intent is to award $200K to $300K a year for a total of $2M. From Harry's perspective, it's important to maintain some flexibility.

Suzan Mazur: Were you involved with selection of the top three proposals?

Lawrence Krauss: I was aware of some of them but stayed away from the selection panel itself. I was very happy to see the prize going to John Sutherland, a member of the prizewinning team who was involved in ASU-OP's origin of life workshop and symposium a year ago.

Suzan Mazur: The exact role you're playing in allocating Lonsdale research funds is what?

Lawrence Krauss: For the moment we'll play an advisory role. We'll see to what extent we can further support Lonsdale prizewinning researchers through events at ASU-OP. John Sutherland will be visiting ASU regularly and we'll be discussing this with him.

Suzan Mazur: Are you and/or ASU-OP being funded by Harry Lonsdale?

Lawrence Krauss: Not at this point directly, although Harry has pledged to make a small gift.

Suzan Mazur: Are you still Associate Director of the Beyond Institute at ASU headed by Paul Davies?

Lawrence Krauss: Yes.

Suzan Mazur: Is ASU-OP carrying out experiments re origin of life or is it an advisory body and conference center?

Lawrence Krauss: ASU-OP tries to enhance ongoing research programs at the university and internationally by bringing scientists to the school and initiating collaborations. ASU-OP is three years old and growing. We just hired our first origins professor. We don't have our own laboratory facilities, *per se*, and I don't envisage ASU-OP having laboratories. With additional funding, ASU-OP will increase support to scientists who do have labs at the university and to researchers elsewhere.

Suzan Mazur: Who is your origins professor?

Lawrence Krauss: Rob Boyd, an evolutionary psychologist and anthropologist from UCLA.

Suzan Mazur: Freeman Dyson said in an appearance on *Charlie Rose* that we are all equally ignorant when it comes to the question of origin of life.

What is your theory of origin of life and how does it fit into your concept of a "universe from nothing"?

Lawrence Krauss: Well that's a big question. Yes, we are all equally ignorant at this point about the exact origin of life. It's an incredibly exciting question, though, and why ASU-OP took an interest in assisting with the Lonsdale prize.

I learned a lot at the symposium ASU-OP had a year ago on origin of life. One of our strengths is bringing people together who don't normally talk to one another. We invited geochemists, geologists, biologists and chemists to the conference.

The interaction of these scientists was fascinating to see regarding issues of importance and standards of proof, like the early climate conditions on Earth and how they might be relevant to the building up of organic molecules -- which is relevant to the work that John Sutherland does.

After that symposium I became convinced we'd come much closer to understanding how chemistry turned into biology.

There's a great intellectual mesh between origin of life and ideas about the origin of the universe. We don't have a complete theory for either one but in both cases we have a plausible set of mechanisms by which we can see a way forward.

With the origin of life, we are actually on the cusp of nailing down that plausible set of mechanisms. It's possible in the next decade or two -- I doubt it will take another century -- we will understand specifically how biology arose from nonbiology. I expect we'll answer the question of life before that of origin of the universe. But in both cases it's a matter of taking what was, before science, considered a miracle and plausibly showing how it arose.

Suzan Mazur: You've said, "God is just an invention of lazy minds." But couldn't the same be said for natural selection? For example, Richard Lewontin noted in the *New York Review of Books* that Darwin intended natural selection as a metaphor, not for generations of scientists to take literally.

Lawrence Krauss: . . . . It's semantics. . . .

Suzan Mazur: Freeman Dyson, during that same *Charlie Rose* broadcast said the following:

> "I like Richard Dawkins as a human being, but I think he's done a lot of harm by telling young people that you have to be an atheist in order to be a scientist. That's a stupid thing to say because it pushes away a lot of young people from science who don't want to give up religion. . . I think religion is not just about belief. It's about a way of life. It's a community. It's a big literature. It's music and architecture. It's a big part of human life which is really not so much dependent on belief. . . Yes, [I am a religious man by that definition] but I certainly don't believe any particular theology."

Would you comment on Dyson's statement?

Lawrence Krauss: Absolutely. First of all, I think he misquotes Richard Dawkins. I have said you don't have to be an atheist to accept science, and the evidence of that is that there are some scientists who are not atheists.

The point that Richard Dawkins would make is -- and Freeman is a friend, so I don't want to put words in his mouth, but I know he must agree -- that science is incompatible with the doctrines of the world's major religions.

Suzan Mazur: How do you view the effectiveness of your proselytizing with Dawkins *et al.*?

Lawrence Krauss: I don't proselytize. I've often disagreed with Richard. One of the biggest impacts I've had is with college audiences and Fox News audiences by saying you don't have to be an atheist to believe in evolution. As Steven Weinberg commented, science does not make it impossible to believe in

God, it just makes it possible to not believe in God. Once you have science, you realize that God becomes unnecessary.

That second part of Freeman's statement is pap. It's nonsense that religion is art and architecture and everything else. It's a part of human culture but the contributions have been far more negative than positive. We could have had art and literature and music without this fairy tale that gets in the way of progress.

Science cannot disprove the idea that there's some purpose to the universe but science is incompatible with the strict doctrine of the world's religions. When you ask people do they believe that a wafer turns into the body of a first century Jew when it's held by a priest, they say absolutely not.

Suzan Mazur: Do you think your vision of a universe from nothing can substitute for what nurtures the religious culture that Dyson describes?

Lawrence Krauss: Well I certainly hope so. I hope that understanding the way the world REALLY works, instead of the way we'd like the world to work, can, if appropriately described, motivate people, inspire them, and help produce the kind of community and society and caring world that actually makes it better than it is when religion plays that role.

For that to happen we have to discuss and understand those aspects of the human psychological condition that religion currently supports. That's the sense of community, and for some people inspiration, which can be richer in a world that accepts the reality of nature and the idea that there may be no purpose to the universe. And that the purpose of our lives is the purpose we make.

I don't think science takes away from a sense of awe, wonder, love and goodness. But as long as we deny reality, we will delay humanity's move forward and away from its current myopic rivalries, hatreds and religious animosities.

Suzan Mazur: Would you comment on why you think the proposal of British chemists, John Sutherland and Matthew Powner, won the Lonsdale prize? Sutherland-Powner will attempt to generate RNA from "feedstock molecules under the presumed environmental conditions of pre-biotic Earth."

Lawrence Krauss: The current best idea is that preceding the DNA world was an RNA world, and that an RNA world could contain much of the metabolism and processors and information storage that eventually the DNA world, which was much more complicated, could house.

The question is, can RNA result naturally? That's been a big stumbling block. The organic chemistry work of John Sutherland is perhaps THE most exciting work in the world, in my opinion, attempting to discover how organic chemistry in the absence of biology can allow the synthesis of molecules that can perform ultimately what RNA does.

For me, it was a revelation to see how far Sutherland had come in terms of trying to understand organic pathways and tell us what precursor molecules might be available either on the Earth or on materials that bombarded the Earth.

Suzan Mazur: The Origin of Life Challenge focus so far has been on the chemical, life described in the Lonsdale outreach as "a self-sustaining chemical system capable of undergoing Darwinian evolution." However, Darwinian and neo-Darwinian theory are waning in influence in serious science circles.

Freeman Dyson gave what he called his "reasonable point of view" about origin of life on that same *Charlie Rose Show*, saying, "You had life evolving without genes for a long time. . ." Microbiologist James Shapiro has described genes as "not a definite entity," they're "hypothetical in nature," saying further that "there are no units" just "systems all the way down." Cell biologist Stuart Newman has written that before there were genetic programs there were physical forces like adhesion, polarity, etc.

Duke University Mechanical Engineer Adrian Bejan thinks origin of life is 100% physics and that both animate and inanimate systems are live systems. Bejan says the natural phenomenon is actually distribution not elimination and that "for a finite system to persist in time (to live) it must evolve in such a way that it provides easier access to the imposed currents that flow through it."

My question is, in giving advice and direction to the Lonsdale Challenge were you aware that whole swaths of the scientific community now have Darwinian science in the margins?

Lawrence Krauss: Well I don't think they have Darwinian science in the margins.

Suzan Mazur: They actually do.

Lawrence Krauss: Hold on. Freeman Dyson is a friend of mine. He was just on our ASU-OP event. We nearly spent a week together and had great discussions about this. Freeman, is, of course, a contrarian and often goes against conventional wisdom. It's a way to provoke people's thinking. Freeman considers metabolism much more important than replication in a sense in the origin of life. And that's an interesting idea.

Suzan Mazur: Bejan and others are taking a physical approach to the origin of life. It's an organized flow of matter and energy. Both animate and inanimate systems are live systems.

Lawrence Krauss: Whether you call it life or not, it's just the laws of physics and chemistry. There's nothing beyond the laws of physics and chemistry that allow for the origin of life. We are just a bunch of chemicals subject to forces and laws. It's electromagnetism and quantum mechanics and how those laws of electromagnetism and quantum mechanics produce chemistry. And how chemistry produces biology. And then once biology is produced, how do those laws impact on how biological molecules evolve. It's a continuum.

To say it's physics is ok, because it is all physics. Ultimately the laws of chemistry are just an application of physics. And so at some fundamental level physics determines the evolution of organic molecules but we call it chemistry because of the way physics manifests itself in the nature of atoms and atomic levels. Biology is an application of chemistry with a certain set of molecules that have certain interesting properties including ultimately to self-organize and produce consciousness and intelligence and affect the planet.

Suzan Mazur: Some say we don't have the tools we need in place to seriously investigate origin of life. What are your thoughts?

Lawrence Krauss: You do what you can do. You make progress and you don't give up. We don't say, "The universe is big, we'll never understand it, let's give up." The point is John Sutherland is demonstrating we have tools to learn more about organic synthesis. You never know until you solve the problem that you have the tools to solve the problem. What you do in science is just try and make incremental improvements and increase your knowledge and get to the goal. You don't know what you need to know until you make that attempt. That statement that we don't have the tools is ludicrous -- we don't know what the tools are.

Suzan Mazur: Do you intend to bring up the Origin of Life Challenge in your upcoming appearance on *Colbert* and will Harry Lonsdale be joining you?

Lawrence Krauss: Harry won't be and I have no idea what we'll we discussing in the *Colbert Report*. I imagine Colbert will decide. I suspect it will be focused more on the origin of the universe and cosmology and the nature of nothing but I'm willing to go wherever he wants to go.

Suzan Mazur: Do you think origin of life is a subject that should be discussed more on television at this point?

Lawrence Krauss: Of course -- the origin of life and the origin of the universe are the two most exciting questions in nature.

Suzan Mazur: Do you lament the lack of forums on public television to discuss this?

Lawrence Krauss: Yes. The problem is TV producers who think people are not interested in science and that you have to dumb down content in order to get people to watch. The reality is the public is fascinated by these open questions.

Chapter 3

David Deamer: First Life &

"The Lonsdale 8"

DAVID DEAMER
(photo, courtesy D. Deamer)

"The Lonsdale Award received 76 entries, many from ISSOL [International Society for the Study of the Origin of Life and Astrobiology Society] members, of which 15 were selected for review. The panel is composed almost entirely of ISSOL members. [ISSOL president] Dave Deamer will chair the panel, and Ken Nealson is vice-chair. The review process will resemble that used in major funding agencies, in which a primary and secondary reviewer read the proposal carefully and report to the full panel during the meeting. This is followed by

a discussion and a confidential vote in which each panelist ranks the proposal from 1 to 5, with 1 being the best score. After the meeting, Mr. Lonsdale will consider the panel scores, comments and recommendations, then choose the winner."
-- ISSOL Newsroom

July 9, 2012

David Deamer's trailblazing of prebiotics goes back to 1977 and a paper on lipids and membranes published in *Nature*, but his curiosity about origin of life dates to his teenage spelunking adventures growing up in Ohio in the 1950s. Then, in 1985, some years after the Murchison meteorite fell in Australia, Deamer began testing organic compounds extracted from the rock by adding water, which caused the material to self-assemble into naturally fluorescent vesicular structures.

Dave Deamer's work is now central to the investigation into origin of life. This past year Deamer served as a principal architect of Harry Lonsdale's Origin of Life Challenge, which announced its results last month with Deamer among the top three award winners. According to ISSOL, the astrobiology society Deamer heads, Deamer chaired Lonsdale's peer review panel that judged the contest. So why would he accept a research award of $60K for his own proposal?

Deamer has told me that he left the room when the paper he collaborated on with NASA's Wenonah Vercoutere came before the panel of peers. Apparently Vercoutere was principal investigator.

Harry Lonsdale confirms that Deamer exited the room and I have not asked for further confirmation from the panel, partly because the names are supposed to be confidential, per the industry practice -- even though Lonsdale's Challenge was a private competition, not federally-funded and bound by the usual secrecy, as evidenced by various leaks on the Internet.

Here for example, in alphabetical order is the probable Lonsdale panel of "**Pros**," otherwise known as "**The Lonsdale 8**":

David Deamer, University of California - Santa Cruz

Jim Ferris, Rennselaer Polytechnic Institute

Peter Gogarten, University of Connecticut

Andrew Knoll, Harvard University

Chris McKay, NASA Ames Research Center

Ken Nealson, University of Southern California

Alan W. Schwartz, Radboud University (The Netherlands)

Jack Szostak, Harvard Medical School/ Massachusetts General Hospital/ Howard Hughes Medical Institute.

David Deamer has further informed me that both he and Harry Lonsdale wanted his and Vercoutere's award run through the competition. It remains unclear exactly why. Much of the award apparently will go to a hands-on research associate.

Deamer's remarks below suggesting that he was on the panel because there is a scarcity of experts in the origin of life field may be an underestimation of the talent out there. It does not exactly mesh with former NASA Astrobiology Institute chief Bruce Runnegar's comment to me for my last book, *The Altenberg 16: An Exposé of the Evolution Industry,* that there are roughly a thousand people affiliated with NAI familiar with these matters.

Dave Deamer, who I also interviewed for my last book, has six patents, and is the author of 126 or so papers and 11 books, among them -- *Being Human: Principles of Human Physiology, The World of the Cell, The Origins of Life: The Central Concepts* (with Jack Szostak), and his most recent -- *First Life*. He is also one of the editors of the MIT book *Protocells.*

His PhD is in physiological chemistry from Ohio State University School of Medicine. Deamer serves on the editorial boards of the *Journal of Bioenergetics and Biomembranes, Astrobiology Journal,* and *Origins of Life and Evolution of the Biosphere.*

My interview with David Deamer follows.

Suzan Mazur: Would you address the perceived conflict of interest regarding your serving on Harry Lonsdale's peer review panel for the Origin of Life Challenge and you and your NASA partner accepting a research award of $60K from that same Origin of Life Challenge -- even though you say you recused yourself as your own proposal was being judged?

David Deamer: This is standard practice in peer review committees that I serve on both at NIH and NASA. The fact is that it's hard for a program manager to find people who do not have some degree of conflict of interest, because we're all interested in the same kind of research and have reached a certain level of expertise in the field. So it's very common for people to leave the room when their proposal is being judged by the peers. I had nothing to do with the peer review of our proposal, nor did I hear about the result until Harry called me later. I had no idea whether or not we were going to be among the winners.

There can be a perception of conflict, I suppose, but it's a practical necessity to bring in people who have expertise in order to judge other grant proposals.

Suzan Mazur: Why didn't you request your research funds be separate from the Challenge, since you advised Harry Lonsdale prior to the contest how to shape the contest? And it would look funny if you received an award within the same contest you were judging. Also, the peers, according to Lonsdale, were not paid.

David Deamer: Yes, I understand that. But, as I say, this is common practice in peer review panels of national funding agencies. The other thing is that this was a unique, privately-funded competition. I wondered about just the question you're bringing up, how would I be able to avoid a conflict. I think if you talk to Harry, he will say that I made every effort to keep apart from any aspect of the evaluation procedure of our proposal.

In terms of the other possibility that Harry would simply fund us separately, he could have made that choice but wanted to run it through the open competition, and so did I.

Suzan Mazur: Since you served on the peer review panel, which, according to Harry Lonsdale, recommended the contest finalists -- can you say why the favored research related to RNA replication, when the consensus among scientists who are thinking about the origin of life question is that something simpler came first?

David Deamer: This was an international competition where literally anyone could propose an idea and a feasible way to test the idea. That's just how science works. People have ideas, you find a way to test them -- if you're an experimentalist -- and you try to find funding to support the research.

So, why RNA compared to all the other possible proposals? Out of the 76 proposals that were submitted, about 15 went through a second selection process with some of Harry's trusted colleagues, and of those the panel decided to review 10 of them. Three happened to be from people working on RNA as a model system.

There's one other thing to keep in mind about how science is done. We need to be practical. The fact is there is a tremendous amount of information about RNA and how it can be handled experimentally. We don't have nearly as much information about other possible replicating systems of molecules that could be experimentally tested. So being practical, a good scientist will choose things that are feasible within their resources and time restraints. It so happened that of the 10 proposals that were reviewed, the three deemed most reasonable were the ones that focused on RNA.

Suzan Mazur: So Sutherland and Powner, their concept of feedstock molecules satisfied the criteria for possibly this simpler form feeding into RNA.

David Deamer: Yes. Sutherland and Powner published a remarkable paper in 2009 showing how these relatively complex nucleotides, the subunits of RNA, could be synthesized by a chemical process using feedstock. Their proposal was not just about RNA but trying to understand the chemistry of systems of molecules that could give rise to a variety of monomers of which the mononucleotides of RNA are one possible outcome.

Suzan Mazur: Did you and the panel have any interest in exploring alternatives to the classic origin of life scenario, say along the lines of animate and inanimate both being live systems, *i.e.*, organized flows of matter and energy, and the universe as a continuum not boxed into living and non-living?

David Deamer: The panel was a group of research scientists who are very well known. These are people who have made real contributions to the field, and they will be looking not just for novel ideas but for ideas that can be feasibly tested within the constraints of the support that Harry is able to provide. For that reason we did not consider pure idea papers. The panel was also looking for ways to test the ideas in order to advance our understanding of the field.

Suzan Mazur: Would you briefly describe your current experiment?

David Deamer: The main idea is that in a solution of monomers, such as monomers of RNA or DNA in solution, the laws of thermodynamics do not allow them to polymerize because there is a tremendous energy barrier to getting them to form bonds. We need to find a way to get them sufficiently concentrated in the absence of water, because water is what breaks these things down. We call that hydrolysis, the same thing that occurs during the digestion of food.

To get water out of the system we use anhydrous conditions. But if we just dry down a solution of potential monomers, it becomes a solid, and diffusion of the reactants cannot occur. Therefore, we have proposed that a liquid crystal can serve as an organizing matrix for the monomers. If we dry monomers in the presence of an organizing matrix, they have an increased chance to form the linkages that allow them to become polymerized. That's the basic idea that we are testing.

Suzan Mazur: You mentioned to me that you don't think your colleagues take your observations seriously. One of those critical of your work was the late Robert Shapiro, who wrote the following in *Nature* shortly before he died last year:

"Unfortunately, his [Deamer's] theory retains the improbable generation of self-replicating polymers such as RNA."

Shapiro wrote further regarding your call "for the construction of a new set of biochemical simulators that match more closely the conditions on the early Earth" that the chemicals you suggest using are, as Shapiro describes them, actually "drawn from modern biology, not from ancient geochemistry." He says, "We should let nature inform us, rather than pasting our ideas onto her." Would you comment?

David Deamer: Yes. Bob Shapiro did a real service to our field by putting on his skeptic's cap and showing us where our ideas are soft. And that's easy to do, because this is a very young field. Only a few hundred people worldwide work in it. And there are lots of ideas floating around, but relatively little experimental work because of the funding situation. Research depends on funding, and NASA is the only source of funding right now for work on the origin of life. It's a limited pool of investigators and a limited pool of money.

The people who are making advances in the field, and I say this once again, are being practical. It is impractical for me as a research scientist to venture off to try a bunch of other ideas, such as Robert Shapiro has suggested. It's not the way I can convince a peer review panel to fund my research -- they would call it a fishing trip. I am trying to be practical about what I do, and RNA is a good place to begin.

Suzan Mazur: Have you gone as far as you can go with what Freeman Dyson calls the "garbage bag world," self-assembled lipid membranes, in terms of hoping to find a copying switch without resorting to mixing in these modern chemicals to force replication?

David Deamer: I worked in Doron Lancet's lab at Weizmann Institute for a while and we later published a paper called "The Lipid World," in which we described a compositional genome as opposed to a linear genome that contains information in polymers like RNA. That is, in fact, a pretty far out idea, but it's fun to think about. We actually tried some of the experiments to test the idea but didn't get very far.

I'm open to the garbage bag idea and compositional genomes, but I also think that life at some point began to use polymers as catalysts and replicating systems of molecules. For that reason, I'm trying to combine the idea of a self-assembled lipid structure -- a boundary layer, or a lipid membrane, if you will -- showing how that can organize the monomers in such a way as to take the next step into the polymers of life.

Suzan Mazur: So you don't have any choice but to mix in the mononucleotides.

David Deamer: It's the best game in town right now. It is.

Suzan Mazur: How do the "feedstock molecules" of Sutherland & Powner differ from your self-assembled molecules?

David Deamer: I'm not sure how John Sutherland and Matthew Powner would define feedstock, but I assume that they mean relatively reactive simple molecules, simpler, in fact, than mononucleotides. However, one end product of the reaction pathways they are investigating is a mononucleotide, as they showed in their *Nature* paper. So if that's what they mean by feedstock, then that's not what my work involves because I don't do that kind of chemistry.

Powner and Sutherland are real chemists, organic chemists. They know a lot about how organic chemicals can behave as a feedstock and produce more complex molecules.

What I'm doing is taking the next step and saying given that more complex molecules can be synthesized, what can we do to get them to form polymers?

Suzan Mazur: Freeman Dyson told me that the garbage bag world scenario probably went on for a billion or more years before replication kicked in. Do you agree?

David Deamer: That's his guess. I'm not even sure I would want to try to guess, but it probably took longer than a week. In the next 10 years of my research career I'm hoping to achieve in the laboratory what we would call a self-assembled replicating system. That would be a convincing version of something emerging from Dyson's garbage bag ideas. At the start of the experiment there's nothing there but a mixture of monomers and lipid, but after we put them through the anhydrous cycling process there are polymers present. The next step is to see whether the polymers can function as catalysts, and perhaps even replicate in some way.

Suzan Mazur: So you don't agree with Freeman Dyson that it will probably take another century before we get to the bottom of origin of life.

David Deamer: **I have heard from Harry Lonsdale that Sutherland thinks it might be 50 years** [emphasis added], and Dyson's guess is 100 years. I don't want to limit it that way, I suppose, because nobody knows for sure. The breakthroughs when they come will be the sort of breakthrough Watson and Crick made when they solved the riddle of DNA replication in just a few years.

Suzan Mazur: Also, if you get more people working on the problem. A concerted effort.

David Deamer: Yes. Rosalind Franklin, of course, provided a very solid experimental test of their ideas. They really were idea guys. They did not test their model except against other people's work. They had a foundation.

Suzan Mazur: What is the importance of computer simulations in studying origin of life?

David Deamer: For one thing they're so much fun to look at. Have you seen the great molecular animations on Jack Szostak's website?

Suzan Mazur: No I haven't viewed those. I'll have a look.

David Deamer: Computer simulations stimulate the imagination. My brain, at least, can visualize things in the simulations that otherwise are just lines and letters on a printed page. Viewing three dimensional models moving around, obeying the laws of chemistry, Newtonian physics and the laws of force fields really does help. I use them in my lectures, they're a tremendous teaching tool.

Suzan Mazur: What value do you see in solving the origin of life mystery aside from satisfying human curiosity?

David Deamer: The easiest answer is spinoffs, if we're talking about applications. For instance, there's already been one spinoff from my work, which is a nanopore sequencing device. It was stimulated by my attempt in 1989 to get ATP [adenosine triphosphate, chemical fuel that cells burn to keep us alive] into a model cell to support a polymerase inside. It occurred to me that maybe ATP would block the ionic current in a detectable way. Then if we pulled a whole string of nucleotides through the pore as a nucleic acid, each nucleotide -- there are only four of them -- would give a differential signal that would let us determine the base sequence of the DNA.

Twenty years later, a sequencing device is going to be marketed by Oxford Nanopore Technologies that resulted from origin of life work research supported by NASA.

Suzan Mazur: What value do you think people in general will derive from knowing the origin of life?

David Deamer: Most people would be interested in a scientific version of how life began. If they're scientists, they'll try it themselves, see if it works.

Suzan Mazur: Is your origin of life investigation also an exercise to dispel current creation myths?

David Deamer: I don't take it that way. It would simply be a scientific explanation that anyone can understand about how life began on a sterile Earth using only available energy sources and organic compounds.

Thoughtful, religious people could still be satisfied in their belief that a supreme intelligence put in place the laws of chemistry and physics. There's no scientific rebuttal to that belief. It would be a valid hypothesis if it could be tested.

Part 2

The Circus Caravan

"In a circus we see mostly what we are ready to see."
-- **Edward Hoagland**, *Balancing Acts*

Chapter 4

Princeton Wagon

The RNA World's Last Hurrah?

ANDREW POHORILLE
(photo, courtesy A. Pohorille)

In a couple of emails to me in January 2013, Andrew Pohorille, the senior-most scientist at NASA working in the origin of life field, objected to my story, "The RNA World's Last Hurrah?" in which I interviewed Paul Davies' collaborator at Arizona State University, physicist Sara Walker. The story title referred to the January 21-24 Princeton University Origins conference that NASA was co-sponsoring and whose presenters were largely working on some aspect of the RNA world. The pure RNA world belief is that RNA (ribonucleic acid) kick-started life.

Pohorille seemed furious regarding story comments doubting the RNA world, although the Walker interview was a Q&A, and I'd quoted the "experts" and linked in my interviews with them: biochemist Pier Luigi Luisi, who characterized the RNA world as a baseless fantasy; theoretical biologist Stu Kauffman, who thought the RNA world hadn't worked; and Walker, who told me "most of the origin of life community don't think that's the definitive answer."

Following are excerpts of my daily reports from Princeton, beginning with my meeting Andrew Pohorille.

January 21, 2013

Last night as the tiny elevator I was riding in opened inside the lobby of the Nassau Hotel, in walked Andrew Pohorille, NASA's senior-most scientist on origin of life. It was just minutes after I'd filed my report with Pohorille's comment that the death of the RNA world -- *i.e.*, life started with RNA (ribonucleic acid) -- had been exaggerated.

By the time I realized who he was, it was too late to introduce myself. The doors had closed behind me, enveloping Pohorille in the scent of Quelques Fleurs Royale. But I bumped into him again this morning in Jadwin Hall at Princeton's origins of life conference, alone, having a quiet coffee, and we struck up a conversation.

Andrew Pohorille has a certain calm and sincerity that draws you in. Born in Poland, he has retained his charming European accent and speaks with an elegant clarity. We talked through some of our points of disagreement. He told me that while he is not an RNA world enthusiast, he does indeed think relevant work is being done on it.

[**Note**: In a subsequent email to me in May 2014, Pohorille wrote that during the Princeton event he had to defend the RNA world, even though he didn't like it or believe in it. At the time of the Princeton conference, Pohorille chaired NASA's Origin of Life Focus Group, which he still does, with Montana State University biochemist John Peters. Pohorille was also point man on creation of the space agency's Astrobiology Roadmap and is currently director of NASA's Center for Computational Astrobiology as well as a professor of chemistry at the University of California, San Francisco.]

We talked a bit about Carl Woese. NASA awarded $8M to Woese and colleagues, three months before Woese died in December 2012, for an investigation into the principles of the origin and evolution of life. I asked Pohorille why he thought Woese had not been honored in his lifetime by the Nobel committee.

Pohorille said it was simply because the great work Woese was doing was outside of the official Nobel categories.

Steve Benner, the synthetic biologist and comedic wit who heads the Florida-based Foundation for Applied Molecular Evolution, was amused by the controversy about the death of the RNA world and said he planned to mention it in his upcoming Princeton talk.

I told Benner, as I had Pohorille, that I'd conferred with the experts on this. Benner said this is origin of life – "there are no experts."

I was then informed that I was being credentialed and admitted inside the meeting. My previous attempts to obtain credentials had been blocked. Princeton is now also streaming the conference over the Internet with the assistance of NASA, which I applaud.

Princeton University biologist Laura Landweber, the event's principal organizer, and I talked at the coffee station with Sara Walker of Arizona State University about the fact that one-third of the presenters at the event are women. Landweber, an accomplished scientist, suggested that next month's origin of life meeting at CERN, called COOL EDGE 2013, should consider Princeton a precedent and invite women to speak. I agreed with Landweber.

Sara Walker gave a commanding presentation on "The algorithmic origins of life" in red stretch jeans and tailored jacket, tossing a marvelous mane of hair and sparking a round of followup questions. One from Nobel laureate Phil Anderson about "Stu Kauffman's point of view that you have to go through energetics first, thermodynamics first and only then. . . ."

Outside the conference room, Anderson told me he was referring to theoretical biologist Stuart Kauffman's ideas about energetics in an article called "Waiting For Carnot," published decades ago in a Santa Fe Institute book. Anderson may have been thinking about Mitchell Waldrop's chapter by that name in his *Complexity* book. In it Waldrop says Kauffman at the time "lacked either the patience or the programming skills to sit in front of a computer screen" to further develop the model.

Also commenting on the Walker lecture was Loren Williams of Georgia Tech, who addressed complexity and the fact that "there are a much higher diversity of molecules in the abiotic world than in the biotic world . . . ."

Kauffman was scheduled to give a talk at Princeton but was unable to travel. He communicated to me that he thinks "organic chemistry, even in space, flows into its Adjacent Possible."

Nicholas Hud, also of Georgia Tech, advised that "we're made of a surprisingly boring number of small molecules, 20 amino acids, etc."

Williams agreed with Hud saying, "Abiotic chemistry makes huge numbers of molecules. It's very sloppy. But biology picked what was useful and excluded all the rest."

Laurie Barge, a colleague of British geochemist Michael Russell at NASA's Jet Propulsion Lab – Caltech, was dazzling in articulating the possibility of alkaline hydrothermal vents on early Earth.

I missed the presentation today by University of Dusseldorf's Bill Martin (a former Russell collaborator) -- "Bringing rocks to life: Hydrothermal vents and microbial origins."

Martin has promised me a private chat. I've been curious about why he left Dallas, Texas for a scientific career in Germany.

"Why did I leave Texas for Germany?" Martin laughed, "A woman, of course."

Tuesday's presentations begin with the much anticipated chemistry of John Sutherland, winner with Matt Powner of the Lonsdale Origin of Life Challenge plus the entertaining Steve Benner on "A semicontinuous process to form oligomeric RNA." Irene Chen of University of California - Santa Barbara, a former postdoc in Nobelist Jack Szostak's lab, will address "RNA fitness landscapes."

But where is the rest of the media?

JOHN SUTHERLAND
(*artist, Peter Sheesley*)

John Sutherland opened day two of the Princeton conference. Sutherland is the British chemist (MRC Laboratory of Molecular Biology - Cambridge) who won Harry Lonsdale's Origin of Life Challenge last year with Matt Powner (University College London). But Sutherland's lecture was so technical that even scientists in the room who are not chemists had a tough time appreciating it. Plus Sutherland was barely audible at times, relating more to the screen than to the audience.

Following the talk, Sutherland abruptly canceled our lunchtime interview and escaped to the airport. So what progress has been made in his investigation stemming from the Lonsdale prize remains unclear, since it was not the subject of his talk. "Reconciling the iron-sulfur and the RNA worlds" was.

Steve Benner, who heads the Foundation for Applied Molecular Evolution successfully resuscitated the audience following

Sutherland's talk. Benner is so far the hands-down star of the Princeton gathering. Wearing an arty teal-blue shirt, he engaged the crowd, punctuating his informative presentation with perfectly-timed jokes while working both the Powerpoint and chalk board.

Benner told me the day before of his plan to open his talk focusing on my story, "The RNA World's Last Hurrah?". And he did.

Italian biochemist Pier Luigi Luisi was not in the Princeton audience to explain why he considers the RNA world a baseless fantasy, as I'd reported. So Benner explained for Luisi, saying the reason why is because ribose is unstable, it decomposes.

Loren Williams of Georgia Tech, who's been a lively questioner at many of the presentations, told Benner the following regarding the RNA world: "Until I see it rain over in a meteor, I'm not going to believe it."

Paul Higgs of Canada's McMaster University, also a recipient of Lonsdale research money, started with a followup to my post quoting Andrew Pohorille's email to me that reports of the death of the RNA world are exaggerated. Higgs said there was indeed progress in the RNA world, citing the work of ISSOL (International Society for the Study of the Origin of Life) president Dave Deamer *et al*.

In explaining how life emerges, Higgs pointed to the importance of spatial effects, saying they're "essential" regarding fluctuations. He said it is fluctuations that take you over the unstable point and life emerges.

Laura Landweber, principal organizer of the Princeton conference, emailed advising the presentation of Georgia Tech's Nick Hud was not to be reported and had not been streamed over the Internet. Apparently, if the details of a scientist's unpublished presentation get out, the science journals won't publish the paper.

As British chemist John Sutherland *et al.* exited following their conference talks and other invited guests were no-shows, the Princeton event further opened up to the public, which invigorated proceedings.

Loren Williams was a morning presenter. Williams heads one of NASA's Georgia Tech teams and addressed the "ironing out of ancient biochemistry," noting that early Earth was rich in iron and "collaborated" with RNA. But Williams said that magnesium was also in the picture, a co-factor in the origin of life, important in autocatalysis.

Nilesh Vaidya, the rising star from Katmandu who's now a researcher at Princeton, agreed with Williams about magnesium's importance in autocatalysis. Vaidya gave the next talk: "Spontaneous network formation among cooperative RNA replicators."

Vaidya told me he has succeeded "at least three times" in creating spontaneously emerging self-reproducing autocatalytic sets from RNA fragments. He did the experiment with Niles Lehman while at Portland State University, prior to his current work at Princeton.

Vaidya thinks recombination could have been the earliest replication tool. In a conversation with me and Dutch computer scientist Wim Hordijk, who also presented on autocatalytic sets, Vaidya said this:

> "Fragments of RNA and lots of magnesium, as you heard from Loren Williams' talk. For RNA catalysis to happen and to bring the RNA fragments together. And some water. Buffer. That's it. RNA starts making one another. That's the unique thing about that system. You don't need to add anything."

[Note 1: It's important to excerpt here my July 2014 interview with Nobel Laureate Jack Szostak concerning autocatalytic sets:

"**Suzan Mazur**: When we met at the Simons Foundation in April, you told me you "don't believe" in autocatalytic sets. Why is that? Haven't the Europeans integrated autocatalytic subsystems into their systems science?

Jack Szostak: Autocatalytic sets is one of those concepts where the people who came up with the original idea, like Stuart Kauffman, rather than admit being wrong, kept changing their story until it was basically the same concept everybody was already working on.

The original idea was that there would be large numbers of compounds where one would help another to replicate, and that one would help some other one to replicate, and that somehow, out of this huge population of interacting molecules, autocatalytic replication would emerge.

In my opinion, that was never chemically realistic. Now you see people talking about non-enzymatic RNA replication and calling that "autocatalytic sets." If that's what you want to call it, that's fine. But it seems like the concept has lost all meaning."]

[**Note 2**: In a conversation shortly after the Princeton conference, Vaidya told me he was no longer working in that area of investigation. It is unclear exactly why.]

Andrew Pohorille discussed the puzzle of functionality of the protein, describing its flexibility and robustness and compared it to a scifi character you just cannot kill. He then showed an image of Tom Cruise dangling from a cliff and quipped that unlike Cruise he did his own stunts.

Loren Williams in a comment from the floor said his observation was that "functional promiscuity went along with structural flexibility" in early proteins. Williams said proteins at first multitask and later become specialized and rigid.

A winter freeze has now set in over Princeton. Conference participants shook off the chill over a group dinner. . . .

January 24, 2013

I was looking forward to ISSOL president Dave Deamer's update on his experiment funded by Harry Lonsdale. Deamer did not disappoint.

In a carefully documented presentation with animation, Deamer discussed how his team formed polymers from monomers. He went from phospholipids to wet-dry cycles. Drying vesicles became multilayered and then the multilayered structures following rehydration reformed into vesicles, capturing inside some of the material that was originally present on the outside.

Said Deamer:

> "What we're going to do in just a few minutes is send xray beams into this kind of a matrix and we're going to see what it looks like when a monomer is organized within the matrix. . . . This is just ordinary decanoic acid, we tried it now with a double-stranded DNA just to demonstrate the power of this. All of these fluorescent vesicles have fluorescent-stained DNA inside the vesicles. So this is what you call a protocell and a potentially replicating molecule captured in membrane survivors. . . ."

Deamer then broke for a commercial message about the machine he invented and its usefulness in pulling RNA through a nanopore. He said it had a 4% rate of error and that the device would be available this year at $900@, joking that everyone in the room should buy one.

I managed to get in a question about how much he was making per machine. Deamer responded that his cut was just a small piece of the action, that the University of California was taking the lion's share.

Andrew Pohorille countered that "a 4% error is terrible, absolutely horrible."

Deamer advised that the company was trying to get the error rate down to 1%.

Pohorille was not satisfied, telling Deamer that was "still terrible."

Pohorille also wanted to know if Deamer was making peptides, saying that if both peptides and nucleic acids could be made using the same general mechanism, that would be significant progress and might be a way around problems origin of life investigators have been "fighting over for years."

Deamer said he's tried a few simple experiments to get peptides but never attempted to activate a peptide.

The origin of life caravan moves on to Geneva and CERN end of February. . .

The Aerialist: Sara Walker

SARA WALKER
(photo, courtesy S.Walker)

"I think we need to move away from treating a strict RNA world scenario as the central accepted answer for the origin of life because most of the origin of life community don't think that's the definitive answer." -- **Sara Imari Walker**

January 16, 2013

Is next week's Origins of Life conference at Princeton University the RNA world's last hurrah? The Origin of Life community has largely rejected the RNA world, biochemist Pier Luigi Luisi recently describing it to me as a baseless fantasy. I asked physicist Sara Walker to weigh in. Walker is on the adjunct faculty at Arizona State University, a NASA postdoc fellow who is one of the presenters at the upcoming Princeton conference.

Long embraced by NASA despite decades of failed experiments, the RNA world is the organizing point for the Princeton gathering, which is co-sponsored by NASA. Walker acknowledges that the origin of life community does not think the RNA world is "the definitive answer." And NASA's award of $8 million in September of last year to Carl Woese *et al.* is proof that origin of life remains largely a philosophical discussion, with Woese telling me in October -- sadly he died in December -- that we don't know what life is, and that his grant to study the principles of the origin and evolution of life signaled that NASA is rethinking its approach to the origin of life problem.

Walker, who is a collaborator of ASU's Beyond Center director, Paul Davies, says she also finds inspiration in the work of Carl Woese and Nigel Goldenfeld regarding collective evolution and horizontal gene transfer along with the ideas of Stuart Kauffman and others on self-organization as an evolutionary process.

Her paper at Princeton is "The Algorithmic Origins of Life," co-authored by Davies.

Walker thinks "biological systems are dictated by the flow of information . . . how information is handled and processed can distinguish living from nonliving."

One of the interesting points of the Princeton conference is the attempt to open it up to parties beyond presenting scientists. A good thing. What's the big secret anyway? That the original RNA world is a bust? That public money has been wasted?

The conference should be wide open to the public, held in a theater like the World Science Festival is, and streamed over the Internet. Public funding might then be a lot easier to come by for origin of life researchers.

Sara Walker's PhD is in physics and astronomy from Dartmouth. Prior to ASU, she was a postdoctoral fellow at NSF/NASA Center for Chemical Evolution at Georgia Institute of Technology.

Some of Sara Walker's professional activities (past and present) include: Administrator, S.A.G.A.N. (Social Action for Grassroots Astrobiology Network); and at Dartmouth College -- Co-founder, Graduate Women in Science & Engineering; Co-organizer, Women in Science Mentoring Program

Excerpts of my interview with Sara Walker follow.

Suzan Mazur: It's good to see that about a third of the presenters are women at the Princeton Origins of Life conference next week, where you are a featured speaker.

Sara Walker: Yes.

Suzan Mazur: You're now collaborating with Paul Davies, director of the Beyond Center at Arizona State University. . . . You've written a paper recently with Paul Davies called "The Algorithmic Origins of Life." Would you establish what you mean by an algorithm?

Sara Walker: An algorithm, in the context of the paper, is a program that allows the active use of information -- information processing -- which is really important to biological systems. It's not just that biological systems store their information in molecules like DNA, but that they actively use this information to operate.

Suzan Mazur: So the algorithm is a program that allows the active use of information. What about the information itself?

Sara Walker: Information can be loosely defined as events that affect and direct the state of a dynamic system. Saying that information is algorithmic really means that specific events are programmed to have specific outcomes in biological systems. So it's really the processing of the information that's unique about how biology operates.

Suzan Mazur: You say that at some point "information gains direct, and context-dependent causal efficacy over the matter it is instantiated in". What is the information you refer to?

Sara Walker: In this case it is the state of the system, an example the connections or topology of a biochemical network, so it is highly distributed. Function arises due to the distribution of information, therefore biologically meaningful information only arises in the context of the wider system.

Suzan Mazur: In your paper you discuss progress being made in understanding where and when origin of life happened. Other scientists I've interviewed differ. Günter von Kiedrowski, for example, has told me the following: "We can't travel back in time, we'll never know the historical course." Steen Rasmussen told me essentially the same. And Doron Lancet said this: "We will likely never know what were actually the exact chemical substances that began life. But we can wisely guess what principles such chemicals had to obey." Would you comment?

Sara Walker: I do agree with those statements. The point is we've made a lot of progress looking at isolated parts of the problem. Identifying what some of the conditions on early Earth were and what you can synthesize under those conditions, what in biology seems to be essential molecules. But to move beyond that and prove an origins story, that's where you really need to get the deeper principles. The examples you gave from von Kiedrowski, Rasmussen and Lancet point to the fact that while we may never know the precise details of the chemistry or the exact sequence of events, we may still figure out the deeper principles at work.

Suzan Mazur: Doron Lancet also told me that we can't even say there were lipids way back when, that what existed might have been lipid-like. And he said it's also very assuming for us to be thinking that the way life is now with 20 amino acids and four nucleotides is "how life should have been from its inception" -- it could have been any set of molecules jump-starting life.

Sara Walker: I totally agree with that.

Suzan Mazur: So have we made progress on when and where it happened if we don't know what the chemicals were and it doesn't matter what the exact circumstances were? Lancet said further:

"[I]f people tell you life began at a temperature of 25

degrees Centigrade, 110 degrees or 360 degrees (in suboceanic vents) -- this doesn't matter. What does matter, is the principle of what would constitute acceptable molecular roots of life, and at the same time have sufficient simplicity to warrant emergence from an abiotic mixture of chemicals."

Sara Walker: We don't even know which chemical systems came first. There's a huge debate between lipids or genetic polymers or peptides. We just don't know, and I agree the best way to find answers is to look at more general principles than precise chemical details.

Suzan Mazur: So all three questions are still up in the air -- when, where and how.

Sara Walker: Yes. We definitely are still up in the air in the origins of life investigation. But one of the reasons I'm optimistic about making headway in uncovering the deeper conceptual principles about origins of life -- the how -- is that scientists are understanding biology better. We're looking at things at a mechanistic level, observing biological systems operate, *i.e.*, how protein networks function, etc. **The problem is that we have to extrapolate from the chemistry and specific details of the life we know to try to figure out the more universal ideas that might be characteristic of any living system including those we haven't identified yet. That's really the hard part.**

Suzan Mazur: Do you agree with the late Carl Woese that our Last Universal Common Ancestor was a process not something material?

Sara Walker: Yes. I'm a very big fan of Carl Woese's work. He's had a lot of brilliant insights into early evolution. . . .

Suzan Mazur: Aside from being a great scientist, he was concerned about ethics in science and spoke his mind. He and Nigel Goldenfeld and their colleagues recently received an $8M NAI grant to investigate the principles of the origin and evolution of life. Woese told me the problem is that we don't know what life is. How do you define life?

Sara Walker: I want to preface by saying that this is an incredibly hard question. And, you need to be careful, it can really bias your perspective depending on how you try and define life.

I have my own working definition that's based on how I approach origins of life, *i.e.*, as a well-defined transition that might be characterized by informational principles. The short of it is that in living systems information has causal efficacy over the matter that it's instantiated in. And it's really this idea that biological systems are dictated by the flow of information. So the hypothesis is that how information is handled and processed can distinguish living from nonliving.

Suzan Mazur: What about the idea of not separating life and non-life? Astrobiologist Maggie Turnbull, for example, says the universe doesn't make the distinction between life and non-life. Carl Woese too talked about putting the organism back into its environment and connecting it to its evolutionary past to again feel "the complex flow." What are your thoughts about this?

Sara Walker: I agree on some level but I think the problem is that when a scientist is trying to work on the origin of life, it's not a well-defined question unless life is distinguished from non-life. Looking at the transition from non-living to living systems you have to define something changing. Therefore, I think it can be incredibly constructive to have conversations about what it is we are saying is changing because it helps in understanding what the processes are governing the emergence of life. It also helps to define what life is, *i.e.*, what we are looking for when we go out to find it in the universe.

Suzan Mazur: Pier Luigi Luisi recently told me the following:

> "The most popular view of origin of life by way of the RNA world, to me and to many others is and always has been a fantasy. . . . This is the story told to students and to me and is something without any scientific ground. . . Until somebody actually finds a way by which randomly produced RNA begins to self-replicate in not one single molecule but in a thermodynamically-driven process, the

RNA world is baseless."

Would you comment, particularly in light of the Princeton conference coming up, where the RNA world will be central to the discussion of origin of life?

Sara Walker: There's a lot of debate about it. The utility of the RNA world has always been that it's easy to study in the laboratory. It certainly seems that there was an early phase in evolution where RNA played a major role in the genetics system. But I don't think we can say anything beyond that. It seems very unrealistic to me that the early phases of evolution would have been dominated by RNA alone. I favor the ideas of self-organizing chemistry.

Suzan Mazur: Metabolism first?

Sara Walker: Yes. I like to think about life in terms of information flows and how information is being processed. And because information is so widely distributed in biological systems, I think there's merit to the idea of autocatalytic sets. Living systems are systems, and we really need to have a systems approach to the origin of life. You can't just start with a single molecule. That's why I like the metabolism-first viewpoint because it really is about how systems act and evolve collectively.
. . .

Suzan Mazur: There a widespread view in the scientific community that we're beating a dead horse with the RNA world. It hasn't worked. Stu Kauffman recently told me why he doesn't think an RNA world can work. Here's what he's written:

> "With the discovery that RNA molecules could catalyze reactions, that is, ribozymes, many biologists became very enthusiastic that the same class of molecules, RNA, could both catalyze reactions and carry genetic information. This is much of what gave birth to the RNA World view of the origin of life. **I find myself deeply puzzled by this, because RNA only carries genetic information via the protein enzymes which properly load each different transfer RNA. RNA carries NO**

genetic information by itself. So I don't see how an RNA world with only RNA can work. . . ."

Would you comment?

Sara Walker: I find it hard to envision how an RNA-only world would operate. . . . Even in modern life, no individual molecular systems replicate independently without interacting with other cellular machinery -- cellular replication is a collective process.

So, I think we need to move away from treating a strict RNA world scenario as the central accepted answer for the origin of life because most of the origin of life community don't think that's the definitive answer.

Suzan Mazur: If the RNA world is the wrong approach, why are so many presenters at the Princeton conference speaking about the RNA world?

Sara Walker: The RNA world is where the experiments have been going on for a long time, and the experiments tell us a lot about molecular evolution. It still has a place in the scientific community.

Suzan Mazur: How do you define evolution? You say "the concept of evolution itself may be in need of revision" and cite Carl Woese and Nigel Goldenfeld. What do you mean by evolution being in need of revision?

Sara Walker: I was thinking about Woese's idea about early life being dominated by horizontal gene transfer, and that life was more of a collective evolutionary process. It's much harder, however, to get your head around the concept of a loose collection, a network evolving. Conceptually, the RNA world is much easier because we can keep imposing the idea of the Darwinian paradigm of an RNA replicator with vertical descent.

Suzan Mazur: But you do think the concept of evolution is in need of revision.

Sara Walker: Yes. I think there are a lot of phenomena in evolution we haven't investigated in as much depth as the standard genetic evolution paradigm.

Suzan Mazur: Would you mention a few?

Sara Walker: Just from this conversation, Carl Woese and Nigel Goldenfeld's ideas about collective evolution and horizontal gene transfer. Stuart Kauffman's pioneering ideas of self-organization as an evolutionary process. The organization really driving it. These concepts are harder to define because they are not vertical.

More generally, in looking at living systems, one of the reasons they are extremely difficult for us to pin down is that the information in the state of the system is a dynamical variable. You can't start with an initial condition and predict where the evolutionary trajectory will end up because the actual way it's changing with time is changing with the way that the system is changing.

Paul Davies and I talk about this. He likes to say that the "dynamical laws are changing with the state of the system" and vice versa, which is a good way to describe it. This is not something that we have a really good way of defining right now.

Suzan Mazur: You say there's a determining factor, it depends on the actual state of the system as to what happens in the future, but Kauffman believes there's no entailing law, that we can't know what's going to happen. Here's your statement: "it is the information that determines the current state and hence the dynamics (and therefore also the future state(s))".

Sara Walker: It depends on how far into the future you're predicting. It has a lot to do with conundrums introduced by self-reference. A great example is M.C. Escher's Drawing Hands, where one hand is drawing another and vice versa. It's difficult to tell what is the cause and what is the effect. Biology is much the same way. For example, we still have a lot of debate about which plays a more dominant role in evolution, genotype or phenotype. The point is that the operation of one subsystem is not well defined without the other, so you can't analyze it as a linear system.

So, I think it's difficult to predict. Kauffman goes so far as to say it would be impossible to predict because the evolution of the biosphere is not an algorithmic process. I can see that as a possibility but I'm not sure.

Suzan Mazur: Would you discuss the distinction between trivial and non-trivial replicators and say why you put lipid composomes in the trivial category when Doron Lancet *et al.* have shown success with lipid composomes with their GARD model, writing that "appreciable selection response was observed for a large portion of the networks simulated."

Sara Walker: Trivial replicators are perfectly capable of evolution. That's actually not the distinction we're trying to make. "Appreciable selection response" means it's an evolvable system.

Suzan Mazur: But you're saying no in the paper, you're putting lipid composomes in the trivial, not evolvable category.

Sara Walker: Trivial/non-trivial is not supposed to be a distinction about evolvable systems. That's one of the distinctions we're trying to make in the paper, that evolution may be necessary to life but it's not necessarily the defining feature.

The trivial/ non-trivial distinction is more about how information is handled and processed in that system. We look at trivial replicators as being systems that can replicate, but they can do so only within the specific confines of their environment and it's really dictated by their environment.

You could have an RNA sequence that could replicate, for example, but could only replicate with the right mixture of nucleotides under the perfect PH conditions, etc. You can contrast that with a living system which has some autonomy from the environment it's sprung from. It can reproduce because of the way it stores information in one place and processes it in another. So it's decoupled, the non-trivial flow of information in living systems gives them some autonomy.

We use von Neumann's example of the self-replicating machine as a simple framework for understanding this concept.

Suzan Mazur: But the von Neumann example, the Universal Constructor, is a machine not a biological system.

Sara Walker: But if you think about von Neumann's idea, the Universal Constructor, which is defined as a machine capable of building any machine in its environment given appropriate instructions including itself, it can provide some insight into how information flows in living systems. The UC requires information stored on a linear tape to tell it how to operate. That information is read out, and based on the instructions read out, specific tasks are executed. The task can include copying the machine.

Biology operates much the same way, except that the information is much more distributed. We like to think all information in biology is stored in DNA but it's much more complicated than that, how the information is read out also depends on the current state of the system, *i.e.*, level of protein expression, environmental conditions, etc. So state of information is important too. The non-trivial information flow makes it a highly nonlinear system.

Suzan Mazur: So you think we need to be spending more time exploring the nonlinear world?

Sara Walker: Yes. I do. I think we need to be exploring much more about how networks process information and how networks evolve and how self-referential systems operate.

Suzan Mazur: Would you like to make a final point?

Sara Walker: Yes. I'm incredibly optimistic about the origin of life investigation. We're understanding these complex issues more and more.

Suzan Mazur: Are you saying we're going to be able to travel back in time or that we're going to be able to make a protocell?

Sara Walker: Neither, but with regard to the protocell, I am hoping we will be making it soon. What I'm saying is that if we can understand the underlying physical principles governing how living systems emerge naturally, that's answering the ultimate question. It's more about the principles, what are the underlying mechanisms?

Suzan Mazur: I have one unrelated question.

Sara Walker: Sure.

Suzan Mazur: People are wondering what happened to Paul Davies' signature moustache. Can you shed some light on the mystery?

Sara Walker: Oh Paul's moustache! It was right before I came to work for him that the moustache disappeared. I think he just got tired of maintaining it. Coincidentally, I had a recent conversation with Paul about going back to the moustache and he said it was just too much work.

Steve Benner, Ringmaster

STEVEN A. BENNER
(*photo, courtesy S.A. Benner*)

"If you don't have a theory of life, you can't find aliens, unless they shoot you with a ray gun." -- **Steve Benner**

Steve Benner describes himself as a "crackpot synthetic biologist to some extent" (he thinks outside the box). Benner pioneered synthetic biology, for example, and is credited with generating the first synthetic gene to encode an enzyme. He is also known for establishing paleomolecular biology involving research into extinct organisms, as well as inventing "dynamic combinatorial chemistry." And he is one of the founders of evolutionary bioinformatics.

Benner currently directs the Florida-based Westheimer Institute at the Foundation for Applied Molecular Evolution. He's also highly entertaining, and if he's ever ousted from the scientific establishment for crackpot science, he could easily replace any of the late night television talk show hosts (except Craig Ferguson).

I first encountered Steve Benner at World Science Festival 2008. He fielded a question I had about Antonio Lima-de-Faria, the 90ish year old cytogeneticist who lives on top of a fiord in Sweden and has vivid ideas about the role of crystals in evolution.

Benner has a way of bringing people in close and lifting them up, describing Lima-de-Faria as "my distinguished colleague from Lund."

He did it again at the January 2013 Princeton Origins of Life conference, this time in an electric talk titled "Is the RNA World Passé?" in which he referred to his "friend Luigi" -- in response to my story questioning whether the Princeton event was the RNA world's last hurrah and citing Pier Luigi Luisi saying that the RNA world was a baseless fantasy. Benner said "Luigi" thinks so because ribose decomposes, it is unstable.

Benner managed to resuscitate the Princeton crowd following the technical presentation by chemist John Sutherland, co-winner with Matt Powner of the Lonsdale prize.

Said Benner, working the room: "All those people who are awake raise their hand." ISSOL president Dave Deamer was first to reach up high, shouting "Yes!"

Benner then moved so rapidly in his delivery that he appeared to be simultaneously at the chalk board and Powerpoint. . .

Steve Benner received his BS and MS degrees from Yale in molecular biophysics and biochemistry the same year, 1976, and in 1979, his PhD in chemistry from Harvard.

His interest in science he said began in childhood -- as a rockhound and later as a pyromaniac. Benner's father was an inventor and engineer, his mother a musician and homemaker.

Steve Benner is himself an inventor, with at least 22 patents. He is the author of the book, *Life, the Universe, and the Scientific*

Method and of 300 scientific papers.

Some of his awards include: Anniversary Prize, Federation of European Biochemical Societies; B.R. Baker Award; Arun Gunthikonda Memorial Award; Sigma Xi Senior Faculty Award.

My interview with Steve Benner follows.

Princeton, January 23, 2013

Steve Benner: I'm running my own non-profit research institute. The Foundation for Applied Molecular Evolution is the corporation in which is embedded the Westheimer Institute for Science and Technology.

Suzan Mazur: I occasionally send you my stories.

Steve Benner: I look at you frequently because you always have a bit of an edge, which I love. You also wrote this book, *The Altenberg 16: An Exposé of the Evolution Industry*, which I thought was entertaining.

Suzan Mazur: I was just stepping into the discussion about origin and evolution of life in 2008. I thought it was important to talk to a diversity of people, including amateurs and mavericks, because who knows where the great idea comes from.

Steve Benner: Exactly. As I mentioned and you've quoted me, none of us are experts and we don't really know what we're doing with this.

Suzan Mazur: Since the Princeton conference is about the RNA world, would you comment as to whether you think things in the RNA world have moved significantly forward in the last few years?

Steve Benner: Yes. They really have. In part, because there's been a swing back and forth between two major themes. One theme emerged in the 1980s, the discovery of catalytic RNA, which led the community, myself included -- I was just a pup at the time -- to think the problem was a lot simpler. The discovery by Sid Altman, Tom Cech, Norm Pace and various people, that there was a potential for RNA being a catalytic molecule was certainly something.

It went back a lot earlier. Carl Woese had this on his mind. You can go all the way back to Alex Rich in 1962. As far as I know, Rich was the first person to actually suggest RNA could perform both the catalytic and genetic roles.

I thought it would be a much easier process. By the 1990s it was quite clear -- as evidenced by Stanley Miller's paper from 1995 or 1996 -- that it was going to be a hard slog to get RNA in its oligomeric form because of the instability of its various pieces. Again, a downer from the 1960s when people were finding prebiotic ways, all sorts of things based on Stanley Miller's earlier work.

Leslie Orgel was one scientist. Juan Oro. All these people found lots of these things going on quite quickly. But in the 1990s we hit another wall with the research. And by the year 2000 scientists were saying -- like my good friend Pier Luigi Luisi, who you've noted has referred to the RNA world as a baseless fantasy. Luigi and I were colleagues at the ETH in Zurich for 10 years, by the way. In the 1995-2000 period, scientists were in despair.

Suzan Mazur: But Luisi is up to date about developments in the RNA world.

Steve Benner: The way he states the case is probably a bit stronger than he actually himself believes. All of us do this in speaking to the Fourth Estate.

Suzan Mazur: He told me something even worse.

Steve Benner: Even worse?!

Suzan Mazur: I asked Luisi if funding was the issue slowing down progress in the origin of life field. He said that is not the problem. Luisi said there are no new ideas, we need "mindstorms."

Steve Benner: There is a shortage of ideas plus there's a shortage of funding to pursue the ideas that we do have. It's an interesting question whether putting more money in will result in more substantial ideas. It's a cycle you can't anticipate, you have to try.

You saw the lecture here at Princeton from John Sutherland, who has lots of ideas. John is trying to explore possibility space. He's reasonably well funded, as British scientists go, because he's at the MRC (Medical Research Council) in Cambridge.

Have you spoken with Matthew Powner, who is a student of his now at Imperial College?

Suzan Mazur: Sutherland left for the airport following his talk, canceling our interview.

Steve Benner: Have you spoken with Powner, his collaborator? Powner is also looking at the process of making DNA.

Suzan Mazur: I have not, although I've mentioned both of them in my reporting, as winners of the Lonsdale prize.

Steve Benner: That's right. They were winners of the Lonsdale Origin of Life Challenge.

Suzan Mazur: But I don't know where they are with that experiment, with that research. That's what I was really hoping to find out from Sutherland.

Steve Benner: I don't know at all.

Suzan Mazur: Sutherland told me he was not talking about it in his Princeton address, but I did think I'd get a little something during our scheduled lunch interview.

Steve Benner: I've been in contact with Harry Lonsdale. He's a chemist. Ran for the US Senate.

Suzan Mazur: Yes. I did an interview with Lonsdale.

Steve Benner: We'd like to host a session with all of the Lonsdale research award winners at the 2014 Gordon Origin of Life conference.

[**Note**: Steve Benner chaired the 2014 Gordon conference. (Chapter 6, "Texas Wagon")]

Two and a half years after the award should be enough time to see some results. I've invited Harry. He said he might come and see what his investment hath wrought.

Suzan Mazur: I would love to attend as well.

Steve Benner: Well you're allowed to come. It's January 12 - 17, 2014 in Galveston, Texas. Again, it's a great place to be because the discussion really is unfettered and unconstrained.

Suzan Mazur: Lonsdale wanted to go really public with the origin of life message. He saw that I'd been on the *Charlie Rose Show* and asked if I could introduce him. I told him what I knew in terms of doing science roundtables these days on the program, which is what Paul Nurse told me in 2008 at his Rockefeller University Evolution symposium. That is, if you want to do a science roundtable on *Charlie Rose*, you need a corporate sponsor like Pzifer, which is who Nurse brought in for his 13-part series.

Steve Benner: I didn't know that.

Suzan Mazur: Some think origin of life is too esoteric a subject for a public discussion, but Princeton and NASA made the wise decision to stream the Princeton conference over the Internet. I think Lonsdale is right. There should be a series of *Charlie Rose* origin of life roundtables.

Steve Benner: I agree.

Suzan Mazur: Luisi says there are no new ideas. Why not widen the discussion circle? Freeman Dyson was on Charlie Rose talking briefly about origin of life. Why not bring Freeman Dyson in again and many others for an in-depth conversation?

Steve Benner: Science, to the public, is at one level the memorization of facts based on an authority -- your teacher, who has the cosmic authority of the expert. You'll see this all over -- "four out of five dentists agree." The appeal to authority and consensus of opinion.

But science is also the opposite. I'm a great fan of Richard Feynman who comments that science begins with a denial of the opinion of experts Science begins when you say NO. The perceived wisdom is wrong. Feynman's opinion is exactly the opposite of what many people think science is, the memorization of facts taught to you by an authority.

Suzan Mazur: That was the view of the late science and technology historian David F. Noble, who told me, "A consensus of scientists. Well, when you have a consensus of scientists, that should set off alarms."

Steve Benner: Scientists must be trained. This is a problem. Feynman goes on to point out that there's an enormous amount of what goes on in the public sphere in term of science that is mostly not scientific at all. He also says there's an enormous amount of intellectual tyranny in the name of science.

Suzan Mazur: Noble thought that peer review was at the heart of the problem. He considered it censorship.

Steve Benner: Well, it's a big problem. I'm a very soft peer reviewer. I think if you want to publish something that makes you look like a fool -- go ahead. Be my guest.

Suzan Mazur: Yes. You have a right to put it up there. Lynn Margulis courageously fought *PNAS* over its system of anonymous peer review.

Steve Benner: Exactly. This is how I don't get asked to review very often.

Suzan Mazur: Some scientists say, well peer review is good because other scientists will add a piece here and there.

Steve Benner: In a good peer review that will happen.

Suzan Mazur: Why not just get the story as right as possible and put it up?

Steve Benner: In point of fact there is some good that comes from peer review.

Suzan Mazur: Plus scientists have to pay to be published. They pay for their articles to appear in peer-reviewed journals. And pay even more if they want the articles open access.

Steve Benner: It's conceivable that the next generation of scientists will go to entirely web-based publication. The reason peer review exists now is that when you print a journal and put it in the mail, you then have a limited resource to meter in some sort of rational way. On the Internet nothing has to be metered.

Suzan Mazur: So is the origin of life field exploding within the scientific community?

Steve Benner: We now have the ability to address most if not all of the 1990s' objections to the RNA world. That means we have ways to get ribose directly. We have ways to get ribose and its nucleotides indirectly. Some of that progress is due to the work of Sanchez and Orgel. Some of these are very old ideas. John Sutherland has amplified these splendidly. There is essentially nothing from the 1995 critique of the RNA world that does not have an answer.

Suzan Mazur: But it will be a creation scenario that COULD have happened. Not what did, in fact, happen.

Steve Benner: There is this general question when the science experiment ends. Peter Galison has written a book, *How Experiments End*. Nothing is ever proven in science. At some point the community decides this is no longer the most pressing problem. That's it. The investigation moves on from this to that.

It's a matter of opinion as to whether the solutions to the 1990s' objections regarding the RNA world are sufficiently robust to allow people to move on from them to something else. It is a little early to say that. Right now what's emerging in the RNA world, which Iren Chen talked about here at Princeton, is that we don't know how useful function is distributed among sequence spaces.

You have 4 raised to the power 100 different sequences of RNA 100 nucleotides long. We don't know how productive function is distributed there compared to destructive function.

One of the things about the model of autocatalysis is that we have many molecules and this increases the chance of having one that does something productively, but it also increases the chance of doing something else destructively.

Suzan Mazur: You said something about straw dog regarding autocatalysis.

Steve Benner: Straw man, I think I said. The modelers had come up with four objections to the model. But none of them are the objections that come to mind first to the chemist . . . The first thing that a chemist says is that if I start adding components to a mixture which is working, the first thing I do is prevent it from working because by adding a component it inhibits an interaction between two spaces. The easiest way a chemist can see this is by crystallization.

If you decide crystallization is what you are going to look for, you go into a laboratory and get an A grade in your organic chemistry laboratory -- as in *Breaking Bad*, the fictional television episode -- trying to do a high-quality methamphetamine product. The first thing you discover is that you must go a long way towards purification of typical organic compounds before they will crystallize. That's because impurities inhibit molecular organization that is necessary for crystallization.

There are a few exceptions. Louis Pasteur is famous for having tripped across one in the tartrate series where you could crystallize spontaneously the left-handed molecule and the right-handed molecule and then pick them apart by tweezers. Very important result in the history of chemistry. If tartaric acid had not been so well behaved, science would have waited another decade or so to get that point clarified.

John Sutherland made a comment about a conglomerate, that is, the failure to spontaneously resolve two-mirror image molecules by crystallization

When you present these models -- say Stuart Kauffman's collectively autocatalytic sets model, one of the earliest, and he's been very influential. I think Stuart would agree with this. The first thing that you see when you're a chemist is an objection. Hey Stuart, you're adding things and thinking it's all good.

Well, very often when you add things, it's bad. It destroys interaction. It inhibits interaction. It catalyzes undesirable side reactions. It depletes material.

Suzan Mazur: Are you involved with making a protocell?

Steve Benner: We're not. There are a lot of very bright people in that business. A lot of them are in Europe.

Suzan Mazur: Are scientific fields here in the US integrating as they are in Europe?

Steve Benner: I saw a piece of yours a while back about the CERN meeting coming up. Partly because of the European funding structure, which we got a taste of in Antonio Lazcano's talk regarding Mexico -- Lazcano said that if you are a professor, you have a chair which has an endowment associated with it so you have a certain amount of research without having to go outside. It makes people lazy and that's one of the criticisms Americans will direct at the European system. But it also permits imagination and broad-based interest.

If you go to France, you see people coming up with cutting-edge advances in sort of this evolution fringe.

Suzan Mazur: This is good.

Steve Benner: Of course it's good.

Suzan Mazur: Well there could be more public interest and more funding in the US. How do you get the public enthused? Do a 13-

part series on origin of life on *Charlie Rose*.

Steve Benner: All of this is coming from Europe. I had 10 years in Europe where I had the ability to do crackpot things. It is because of the funding structure in Europe. I worked for 10 years in Switzerland at the Swiss Federal Institute of Technology in Zurich (ETHZ). I was the beneficiary of the European funding system -- for which I'm eternally grateful -- out of which major advances came, which would not have been conceivable at all in the United States.

You heard Eric Goucher's talk here at Princeton, where he was resurrecting ancient enzymes. We were the first people to do that. We did that in Europe. We would never have done that in the US. I proposed that to the NIH when I was just a pup in the mid-1980s, but the NIH panelist reviews were scathing. And I've kept them.

The second thing is that we reinvented the genetic alphabet. Put 12 letters into DNA and the four ACTG. But go around the world and ask where has synthetic biology expanded on the notion of what might constitute a genetic system. Well in Belgium there's Peter Herdwijn. You go to Albert Eschenmoser at ETHZ, my former laboratory. This is all coming from Europe. There is nothing of this power originating in the United States.

Suzan Mazur: But you're a synthetic biologist now in America.

Steve Benner: That's right. A crackpot synthetic biologist to some extent. Synthetic biology is now mainstream. It was certainly not mainstream in the early 1980s. Or in the 1990s when Pete Herdwijn was getting started in Belgium.

It was only because of the funding structure of Europe where you were allowed to bring together -- you were mentioning this in the protocell area, but this is 20 years ago -- the molecular biology needed to do this, the synthetic organic chemistry needed to do this, the physical chemistry, the biophysics, this concept about origins of life to inspire. You didn't see that anywhere in the United States until after Europe took the lead. That is a direct assignment, you can attribute that directly to funding.

Nick Hud, the Lion Tamer

NICHOLAS V. HUD
(*photo by Mona Hud, courtesy N.V. Hud*)

I decided science had come a long way from its roots in misogyny after listening to a recent talk by Georgia Institute of Technology chemist Nicholas Hud in which Hud described his mother to an Atlanta Science Festival gathering as "more important than science," saying further: "She's really shaped me to be who I am more than anyone else . . . she is one of the people I most admire in the world."

It reminded me of the admission by a man writing in to the *Financial Times* a few years ago about how much he enjoyed golfing with his wife -- which the *FT* found irresistible and featured as a Letter to the Editor.

Hud was clearly a wise choice for director of Georgia Tech's Center for Chemical Evolution (CCE), which serves as the hub for 20 laboratories around the country investigating origin of life and is funded by NASA and the National Science Foundation. Not just because Hud is a superb scientist who likes women (his wife Mona took the above photo for his official George Tech page), but because he's also able to bridge the worlds of science and religion. As a youth, Hud spent 16 years in Catholic schools. He knows how to educate audiences about origin of life without eviscerating belief systems and losing the crowd.

Over the last decade, Nick Hud has been researching precursor molecules to RNA (ribonucleic acid), which he calls proto-RNA: "I'm trying to find the simplest molecules that would have been able to take the first steps toward what was needed to start life."

I first met Nick Hud at the 2013 Princeton Origins of Life conference where his talk "RNA evolution and my grandfather's axe" was not for distribution because his work was as yet unpublished.

More recently Hud has popped up on Harry Lonsdale's Origin of Life Challenge web page giving valuable comment on the progress of Londale's various origin of life research teams, and he's been profiled in the Simons Foundation's magazine *Quanta*.

Aside from serving as director of Georgia Tech's CCE, Hud is currently associate director of the university's Parker H. Petit Institute of Bioengineering and Bioscience. He is also involved in research at the school's centers for drug design and delivery as well as nanobiology of macromolecular assembly disorders.

Nick Hud's PhD is in biophysics from the University of California, Davis. He was a postdoc at Lawrence Livermore National Laboratory and a National Institutes of Health postdoc at the University of California, Los Angeles.

Hud has been a visiting professor of chemistry at the National NMR Center in Ljubljana, Slovenia; Imperial College London; and in Barcelona at the Spanish National Research Council and Institute for Research in Biomedicine.

I spoke with Nick Hud by phone at his Georgia Tech lab. Our interview follows.

September 25, 2014

Suzan Mazur: Your PhD is in biophysics but you consider yourself a chemist.

Nick Hud: My undergraduate degree is in physics and my research in graduate school was in biophysics. Biophysics really brings you over towards chemistry. When you get down to looking at molecular interactions, it really becomes chemistry.

Suzan Mazur: That's a wonderful range. You're now also director of the NSF/NASA Center for Chemical Evolution at Georgia Tech involving 20 labs around the country. What does your center do in terms of bringing those labs together, as a hub for all that activity?

Nick Hud: We are a highly integrated effort to solve a chemical problem that is key to understanding the origin and evolution of life. We have within the center joint research projects, so we're not just talking to one another and collecting information. We're collaborating. We have jointly advised students and postdocs and we freely exchange information among our labs. We have projects with meetings, sometimes once a week, where we discuss progress made in the different labs as we're pursuing those projects. We have center-wide meetings once a month where two students or postdocs share their progress. Once a year everybody gets together and we all exchange information. It's a megacollaboration on this one project.

Suzan Mazur: Plus you have your own research. Can you tell me briefly where you are in your research on "proto-RNA," what you consider the first biological molecule?

Nick Hud: As you know, RNA is a complicated molecule. RNA has nucleobases in its center that pair. A sugar, ribose, is attached to the base and phosphates link the nucleotides.

So when I look at RNA, I think of it as three different molecules that have been joined, because the chemistry that would give rise to the nucleobases is pretty different from the chemistry you might expect to give rise to the sugars. And they're both different from

86

the phosphate, which is an inorganic molecule rather than organic molecule.

For years we have been working to understand how a molecule such as RNA could have arisen without the aid of enzymes and without the aid of the chemistry that we can do in a modern laboratory. It's something that would have happened on the pre-biotic Earth.

The hypothesis that we've been working on is not an original hypothesis. It's something that was first discussed years ago, even decades ago, by people like Stanley Miller and Leslie Orgel. It's the idea that RNA could be the product of evolution. We think that each part of RNA may have changed over time, since the first proto-RNA was formed.

We're looking at each of the components of RNA and asking the question: What could have come before this component? If we make a change chemically, if we work with slightly different molecules in the lab -- how could that facilitate the spontaneous formation of this molecule?

Suzan Mazur: I saw a comment by Michael Yarus that you're having a difficult time with making proto-RNA because the bonds break easily. It falls apart. And RNA itself falls apart.

Nick Hud: I think that comment was in reference to what we might even call a pre-proto-RNA. In the work that Michael Yarus was commenting on, we demonstrated two important results with molecules that are very much like the bases found in RNA today, but different. When we can add these molecules to a solution that has sugar molecules in it, they spontaneously form a chemical bond with the sugar that makes them nucleosides -- which is a real stickler of a bond to make with the bases of RNA. These molecules easily make that bond, and then they self-assemble. These proto-nucleosides come together into assemblies that have paired bases that are stacked into long fibers.

What Mike Yarus has said as a criticism is that these proto-nucleosides are not yet linked to each other by covalent bonds. He and I have had discussed this point.

He says, "Well, you don't have information. You don't really have an RNA polymer."

Absolutely. We don't have that yet. What we have is a molecular assembly that is perhaps one bond away from making it a polymer -- a covalent polymer. We're looking for mechanisms, chemistry that would allow the molecules to assemble in water, in solution, and then the chemistry reactions will stitch them together.

So I agree with the Yarus claim that we're not there yet, for sure, but I don't think that there's a problem with our system. It's just that we haven't yet got the chemistry to the next step, but we're working on it.

Suzan Mazur: Obviously, you think the RNA world model is worth exploring. People like Pier Luigi Luisi continue to be negative about it, saying experimenting in the RNA world is like trying to build from the roof of the house. But you think your approach can really provide the basement to the house.

Nick Hud: You have to look at the RNA world hypothesis in different ways. As you well know, there are different models that people are speaking of when they talk about the RNA world. There's a range.

There's a strong RNA model where RNA is considered to be the first molecule of life. In this statement of the hypothesis, RNA makes all the first enzymes and invents everything necessary to start life. Then there are the less extreme RNA world models, where people say that RNA was once more important than it is today. They say that some biological mechanisms that RNA used to take care of -- different catalytic activities -- have been taken over by protein enzymes, but RNA was not necessarily the first polymer of life. What we're looking at is what came before RNA.

Suzan Mazur: So you're building from the basement.

Nick Hud: I'm trying to find the simplest molecules that would have been able to take the first steps toward what was needed to start life.

Suzan Mazur: I'd like to ask you about a comment you made about the autocatalytic sets research of Niles Lehman and his collaborators in a video Harry Lonsdale posted online for the Origin of Life Challenge's annual review of projects. Lehman *et al.* are trying to isolate an autocatalytic set.

You made a comment that their experiment showing the stitching together of RNA fragments had to do with the introduction of manganese. Can you say more about that, in light of Jack Szostak's recent comments to me that he "doesn't believe" in autocatalytic sets and he doesn't think the autocatalytic sets model was ever really "chemically realistic." That the whole model has become meaningless because it just morphed into non-enzymatic RNA replication, which everybody had already been working on. What are your further thoughts?

Nick Hud: I also have a problem with autocatalytic cycles -- with the chemistry, the utility and the wide range of what people are talking about. There is a long history of people proposing autocatalytic cycles and thinking that such cycles are central to starting life, but then not getting much to work in the lab.

I am a great fan of the idea of cycles, but I mostly think of cycles that are driven externally, such as day-night cycles that are driving chemicals between different states. To me, that seems a lot more reasonable as the type of cycle that could get life started.

When it comes to an autocatalytic cycle where you just put the molecules into a solution and they replicate, they have to be very specialized molecules. Only certain ones will do, and a lot of engineering goes into making these.

Whereas, we have already demonstrated that you can take some very simple molecules and subject them to wet-dry cycles, and you can make polymers. These polymers break apart and they reform. Those are the type of reactions that I think you need to get life going. You can call that a cycle -- a cycle of making and breaking bonds. But you don't need very special molecules, aside from the fact that they're stable and they have certain reactive groups that react together at mild enough temperatures.

Suzan Mazur: What do you think the significance is of the current investigation into "making matter come alive"? No genetic material involved.

Nick Hud: I've been to meetings where they've been talking about things like that. . . . I've spoken with Lee Cronin at the University of Glasgow, for instance, about his approach versus what we're trying to do. I think what they do is interesting and that in some ways they'll get parts to work more easily, more robustly than we can at the moment. If you just want to do chemistry, you could have some structural entity that can replicate and undergo exponential growth -- they can do that. But one of the issues is that their molecules are limited with respect to structure.

If you have an inorganic molecule, how many different variations can you have of it? This is where I think organic polymers -- like polymers that we have in life, or artificial polymers even, that have different ways that you can arrange monomer units -- allow you such a great versatility in the structures and functions that you can have, that I think those are ultimately going to be the type of systems with which we will be able to demonstrate artificial evolution with.

Suzan Mazur: The other thing I wonder about is that Cronin's approach is "survival of the fittest," implying Darwinian evolution. But it is now thought early life had a different kind of transfer of information, lateral or every which way.

Nick Hud: This is where we get into some subtleties, because I think that survival of the fittest doesn't necessarily mean Darwinian evolution. I've been thinking a lot about this too.

If you look at a paper by Carl Woese published back in 2002 in *PNAS* (*Proceedings of the National Academy of Sciences*) where he talked about the origin of cells -- a wonderful paper -- you will see that Woese proposed some type of chemical evolution before life crossed over what he called the Darwinian threshold.

Suzan Mazur: I interviewed Carl Woese and Nigel Goldenfeld about this. In fact, in my recent interview with Nigel Goldenfeld, he pointed out that most life is microbial and operates not with a

90

vertical transfer of information but, as Woese said some time ago, horizontal. Woese thought these early cells exchanged information by just smashing into one another.

Nick Hud: But just because you can accept that life is not limited to vertical gene transfer, it doesn't mean that you have to rule out survival of the fittest at the earliest stages. What it comes down to is that we don't seem to have the selfish gene early on.

Suzan Mazur: From what I gather -- the move in understanding life is away from gene-centered thinking about evolution. So the question is what kind of transfer of information is going on, even now? You've got the return of Lamarckian science coming on strong. Our own microbiota inside and outside of us are operating by lateral transfer of information. All very fascinating. . .

What do you think about the current DOE investigation at Fermi Lab into whether we live in a 2D world or a 3D world? Fermi is using its Holometer with twin lasers to try and figure out if we're living in a hologram, whether the universe "could actually be encoded in tiny packets in two dimensions."

Nick Hud: I haven't looked into that. I'm not sure why that would be the case.

Suzan Mazur: Do you think the origin of life investigation is exploding or imploding? Labs like Luigi Luisi's and Norm Packard's have lost their funding for synthetic cell development, for instance. On the other hand, the Simons Foundation is supporting an origins collaboration and Lonsdale has refunded his teams for another year.

Nick Hud: I think it's changing. I think it's an exciting time. I think there are actually quite a few new ideas out there. I wouldn't say that it's exploding or imploding, I think it's just changing. It's never been a field where there was a lot of money for research. It's always been a smaller enterprise than say cancer research, for understandable reasons. I don't know if it's changing in the bulk amount of research, or if there are just shifting trends in focus.

Suzan Mazur: But as you pointed out in the Lonsdale video, Wall Street is now indirectly supporting the research. Dimitar Sasselov told me his group is now looking for corporate investment.

Nick Hud: That would be terrific. There's philanthropy, there are people who want to see this done, who see the promise in it -- longer term. I think corporations look to invest with shorter-term interest, and that's understandable. There are practical applications. It would be wonderful if corporate interests could appreciate that some of the research is now close enough and could pay off.

Suzan Mazur: What is promising about the research? Why would corporations now want to invest?

Nick Hud: If a group or a number of groups were very close to making a system that could self-assemble and evolve and develop a useful function -- then your investment could allow you to own that technology. I think that could be worth a lot. I'm a perennial optimist on this.

Suzan Mazur: Drug delivery? What applications do you envision?

Nick Hud: Drug delivery. It could also be for new biocompatible materials. It could be for new catalysts. It could be for what we call smart materials that change depending on what stimuli they're given.

These are different things that we're only able to do right now with biomolecules. For example, the beautiful work led by Jack Szostak showed how the directed evolution of molecules is incredibly powerful, and you can do it very well with RNA. You can pull out RNA molecules that are enzymes.

Imagine if we could do that with non-natural polymers that have different properties. As you know, RNA is not the most stable molecule. Imagine if you could do directed evolution with something that was much more stable, maybe even more versatile. Maybe you could do it without enzymes, as you need now for RNA evolution. Maybe it will be as simple as drying molecules down and then rehydrating them, or subjecting them to multiple wet-dry cycles.

And maybe you could do it with inexpensive materials. Again, I'm the optimist on this, but I think it will be possible in the future to make molecules cheaply with which we can evolve new functions. I think the origin of life was that way -- simple. I think the molecules that got it all going were very robust molecules. I don't think there was anything special about them, so to speak, with respect to their synthesis. They are special in what they can do. When this problem is solved, I think it's going to be possible to buy the ingredients really cheap and then just cook them up to make useful materials.

Suzan Mazur: Do you consider what you're doing in your lab as protocell development at this point?

Nick Hud: No, we're really looking at proto-RNA. Proto-nucleic acids. We're also doing some work on proto-peptides. It's really proto-biopolymers.

Suzan Mazur: What do you think about Szostak's statement that he hopes to have life in the lab in three years?

Nick Hud: Some of it's going to come down to what your definition of life is. It may come down to, yes, you have something interesting, but is it alive?

Loren Williams:

"The Original RNA World Model Is Dead"

LOREN DEAN WILLIAMS
(*photo, courtesy L.D. Williams*)

"We know the abiotic world is very good at making amino acids. Abiotically produced amino acids are literally raining down on our heads, in meteors. It is difficult to imagine a primitive biological world of evolution, ultimately producing RNA and RNA precursors, that did not make use of amino acids, which were highly available from the get-go, and are so easy to combine to form wonder polymers. So the pure RNA World -- I don't accept it. There had to be many other molecules involved. And a dirty RNA World is a co-evolution world, with protein and RNA evolving together." -- **Loren Williams**

Loren Williams is a professor of biochemistry at Georgia Institute of Technology. He was one of the liveliest questioners at the January 2013 origins conference at Princeton -- where we had a chance to speak -- consistently pitching with a relaxed focus. Williams was having fun. He was a natural, and I was not surprised to learn that he was born into a science family.

Williams' roots are in Seattle, Washington. His father is a marine biologist, his mother has a degree in microbiology. "It was one big science project all the time," Williams said laughing.

He is now also principal investigator of Georgia Tech's NASA Astrobiology Institute-funded center. Williams says his lab is looking to chemically rewind the "tape of life" to before the Last Universal Common Ancestor (LUCA) of organisms and is studying "the transition from nucleic acid-based life to protein-based life." He does not think we got to RNA abiotically, that is, he thinks it happened through biology.

Loren Williams has a PhD in physical chemistry from Duke University. He was a postdoctoral fellow at Harvard, a National Institutes of Health postdoctoral fellow at MIT and a postdoctoral associate at Duke University. He's been a visiting professor to both the University of Sheffield, UK and the Tata Institute of Fundamental Research, India.

Williams is the recipient of an NSF Career Award (1995-1998), Sigma Xi Award (2009), and Petit Institute's "Above and Beyond" Award, among other honors.

He is currently a member of the Executive Committee of the NASA Origin of Life Focus Group, has served on NASA's Exobiology Review Panel, and assisted with the development of NASA's Digital Learning Network.

Our interview follows.

Princeton, January 23, 2013

Suzan Mazur: Have we seen significant developments in the RNA world in recent years?

Loren Williams: The RNA World is evolving. If you look at how people presented it in the past compared to now, there have been significant changes. The 1986 paper by Wally Gilbert describes a pure RNA world, where you have RNA running all of metabolism and information transduction. It's all RNA, all the time, and nothing else. I don't think many people accept that as reasonable anymore. In that sense the original RNA World model is dead.

But what has happened is that people have adapted and modified the model. There are now dirty RNA World models in which RNA collaborates with other molecules. Most people now are assuming that amino acids, peptides and a variety of other molecules were involved along with RNA.

Suzan Mazur: So it's kind of what John Sutherland's been saying, it's important to integrate.

Loren Williams: Yes. That is definitely happening.

Suzan Mazur: Is it now happening in the US as it is in Europe where you have different fields coming together?

Loren Williams: I think it's a global phenomenon. And I wouldn't call the RNA World a dead end. Most people would say it's a powerful model in part because it makes very specific predictions that can be tested. By testing those predictions it can be seen that the original RNA World has problems. The RNA World predicts that RNA can efficiently polymerize RNA, which seems problematic, biochemically. Another problem is related to the origin of RNA itself. Where did RNA come from? Where did RNA precursors come from?

Steve Benner and John Sutherland are working on trying to understand abiotic routes for production of RNA precursors.

However, Nick Hud just gave a really beautiful talk here at Princeton in which he showed data suggesting that RNA is a product of biology. In Hud's model there was an evolution of polymers -- RNA was not the first informational polymer, but is the winner of an evolutionary process. To me this makes perfect sense. But broadly speaking, there are a variety of possible solutions on the horizon to the question of where RNA came from. Benner, Sutherland and Hud will work it out.

Benner and Sutherland hypothesize that abiotic systems made RNA precursors. Steve Benner suggests borate as a co-factor. Nick Hud would say no, RNA is a product of biology and you're never going to find an abiotic process to produce RNA or RNA precursors -- it didn't happen that way.

The question of the origins of RNA is closely related to the nature of the RNA World. We know the abiotic world is very good at making amino acids. Abiotically produced amino acids are literally raining down on our heads, in meteors. It is difficult to imagine a primitive biological world of evolution ultimately producing RNA and RNA precursors that did not make use of amino acids, which were highly available from the get-go, and are so easy to combine to form wonder polymers.

So the pure RNA World -- I don't accept it. There had to be many other molecules involved. And a dirty RNA World is a co-evolution world, with protein and RNA evolving together.

Suzan Mazur: Do you think interest in the origin of life field is exploding or is it relatively static and viewed as esoteric?

Loren Williams: It has exploded for me! The public is interested, people are very interested. Origin of life draws people to science.

Suzan Mazur: And the scientific community? Because I'm hearing a diversity of opinion on this.

Loren Williams: I think things are changing there. There used to be a sense that you shouldn't study origin of life because there was not a lot of funding for it and the questions were not answerable. For a young person to embark on origin of life research was considered professionally dangerous.

Suzan Mazur: It's not completely without clues. There's something tangible there.

Loren Williams: You're absolutely right. Yes, you're absolutely right. It is a solvable problem. It's a difficult problem. It's also not a binary thing where it's solved or not solved. It's going to be a long continuum. The solution to understanding the origin of life will never be a fully finished product. We understand much more about the origin of life now than we did 15 years ago. And we'll understand much more in 15 years than we do now.

Suzan Mazur: Are there more conferences coming up?

Loren Williams: There are lots of conferences coming up. One thing is that funding changed so much. It's so much harder to get funding now for anything.

Suzan Mazur: For origin of life research?

Loren Williams: For anything.

Suzan Mazur: It appears that funding for origin of life research is increasing.

Loren Williams: Relatively, that's right. Compared to other things origin of life doesn't look so bad anymore.

Suzan Mazur: It's coming from the private sector.

Loren Williams: And from NASA and the NSF, there's money for origin of life. People should realize that funding for science, especially basic science is drying up. The NIH is funding grants at less than 10%. So the idea that it's foolish to study origin of life because you can't get funded is not as compelling an argument as it used to be.

Suzan Mazur: My understanding is that Templeton funding is a pain in the neck to try to get.

Loren Williams: That's small money, really.

Suzan Mazur: Some people are getting a fair amount of money but the process is long and demanding. . . . One Templeton grant applicant I spoke with, a distinguished author, told me they found it necessary to hire a professional grant writer for $10K just to complete the application.

Loren Williams: All scientists spend a lot of time preparing grant proposals, most of which are denied. That is our life. Templeton is not like NSF or NASA's Exobiology or Astrobiology programs. Those are much bigger, more professional programs that are engaged in the sustained effort that is required. If the federal government isn't going to fund it, it's not going to get done. At Georgia Tech, between my NASA-funded center and Nick Hud's NSF-funded center, we may have more people working on origin of life than anywhere in the country.

Origin of life is not independent of other areas of investigation. I can tell you that in my research, we are studying the ribosome and the role of the ribosome in ancient biology and the origin of life. What we're learning about the ribosome is going to allow us, I believe, to search more efficiently for antibiotics. Research in biomedicine and origin of life is intertwined.

Carl Woese is a good example. He was interested in the deepest and most profound questions in biology. His discovery of a third branch of the tree of life obviously has profound implications for human health.

Suzan Mazur: Some people would go further. James Shapiro told me he considered Carl Woese the most important scientist of the last century.

Loren Williams: As a chemist I was taught to reserve that title for Linus Pauling. But Woese, well I am siding with Shapiro on this one. Carl Woese used to make the argument that you can't understand cancer without understanding deep biology. You can't understand things in isolation. And he rewrote the book of biology for all of us. Who can argue?

Suzan Mazur: I did probably the last feature interview with Carl Woese. Woese wanted to put the organism back into its environment and see the complex flow. It seemed almost like he was saying you can't separate life from nonlife.

There was all this discussion in the Princeton talks about LUCA as something material, but Woese was not thinking like this toward the end of his life. He saw LUCA as a process, not something material.

Loren Williams: But it was a population rapidly exchanging genetic components, a diverse population and -- this is my interpretation of his work -- as complexity rose the exchange levels slowed down. The exchange level got low enough and speciation took off.

Suzan Mazur: But toward the end of his life he was not thinking of LUCA as anything material. He was thinking about LUCA as a process. . . .

Loren Williams: I don't know if I'm there.

Suzan Mazur: In my interview with Woese, he told me we don't know what life is.

Loren Williams: I agree. People talk about how we define life. You can't -- well I can't.

Suzan Mazur: That's what he was investigating with his NASA research grant.

Loren Williams: Yes. Life is complicated chemistry.

Suzan Mazur: Woese was seeing it more as physical processes.

Loren Williams: If you were sitting in a plastic bubble, watching the origin of life on Earth -- you'd never be able to say: Oh snap, life just started. There is no bright line. There is a smooth continuum between general chemistry and biochemistry.

Suzan Mazur: If you go to the website of Woese's collaborator Nigel Goldenfeld, you see this time lapse of formations in Yellowstone Park. What happens over a period of time, studying the processes.

But you think the public is interested in origin of life.

Loren Williams: Yes.

Suzan Mazur: NASA should help fund a series of roundtables on

Charlie Rose. Get the message out there.

Loren Williams: I think they should.

Suzan Mazur: Who knows what ideas are out there, beyond the scientific establishment.

Loren Williams: It's a great way to sell science to the public. . . .

Suzan Mazur: So you don't think origin of life is too esoteric a subject.

Loren Williams: No, it's not too esoteric. And I think you have to give NASA a lot of credit for its outreach. Other federal agencies don't do such a good a job at trying to foster public interest. The NSF tries, but they don't get to go to Mars.

Suzan Mazur: It's wonderful that Princeton and NASA streamed their origins of life conference over the Internet.

Loren Williams: Yes. And origin of life is a solvable problem. Science, especially chemistry, is by nature indirect and inferential. Everything we know about chemistry we infer through indirect methods.

Suzan Mazur: Günter von Kiedrowski, considered by some the inventor of systems chemistry, told me "we can't go back in time," that we'll never know the exact origin of life, the "historical course."

Loren Williams: That's BS. We can't see atoms with our eyes, yet we know there are atoms. We didn't witness the formation of the Grand Canyon. We have not seen a live dinosaur. We use inference. I am sorry, but "we can't go back in time" is nonsense. I guess the modifier 'exact' gives him an out.

Suzan Mazur: You mentioned something earlier about Stu Kauffman and autocatalytic sets. . .

Loren Williams: Well, I want to be careful not to offend anyone. But please understand that I am a simple-minded biochemist. When theorists and physicists go off on emergence, complexity and coherence, I always say (very quietly) 'show me the molecules'. We have very powerful and well-developed concepts like free energy (G), enthalpy (H) and entropy (S). Those concepts explain and predict. Emergence, complexity and coherence don't mean anything to me.

Chapter 5

CERN Wagon

Stuart Kauffman: Waiting for Ganesh

STUART KAUFFMAN
(photo by Suzan Mazur)

There have been obstacles for theoretical biologist Stuart Kauffman *et al.* in realizing an origin of life brand with CERN (Conseil Européen pour la Recherche Nucléaire). Obstacles that so far even Ganesh, the Hindu elephant god Kauffman adores, has not been able to surmount. As I write, October 2014, the dream is on hold.

The first workshop about a possible collaboration was hosted by CERN in 2011 at supercollider headquarters outside Geneva on the Swiss-French border, with some of the world's most extraordinary scientists participating in discussions. But after more than three years and two strategic meetings, no one has been willing to dedicate themselves to the nurturing of Kauffman's origin of life brainchild.

The last meeting, COOL EDGE 2013, was termed a "fiasco" by the donor who provided funds to the group until promised Templeton Foundation money kicked in to cover travel expenses, because there has been no forward movement. Some conference participants I spoke with recently voiced similar dissatisfaction, one saying the idea was "unworkable from the start." Templeton has not made public its assessment.

Funds for COOL EDGE 2013 had to be channeled through the University of Vermont because CERN will not accept money from private sources. Harry Lonsdale's Origin of Life Challenge money was similarly run through Arizona State University. This raises a whole other issue about universities acting as bankers, taking a fee and having to hire personnel to administer funds, because it inflates the cost of a college education.

Scientists are among the best paid intellectuals -- at least in America. They share in the profits earned from their patents and book deals aside from substantial institutional salaries. Some are independently wealthy. So, it is unclear why the origin of life scientists meeting at CERN didn't just dip into their own pockets for plane tickets if they believed in the mission. CERN had already agreed to provide on-site accommodations.

Some COOL EDGE 2013 participants ultimately seemed unreceptive to the arrangement in which they'd eventually have to find money for the experiment, since CERN will help develop projects but won't fund them.

CERN helped launch the World Wide Web, for instance. Its organizational model provides for a meeting place for such experiments, use of CERN's other facilities and lending of CERN's name to the collaboration.

The current lack of momentum does not mean the origin of life project can't happen someday. Indeed, Markus Nordberg, CERN's director of Resources Development for Development and

Innovation, remains optimistic that it will get done -- even if it takes a decade and the torch gets passed to a new generation of scientists. In fact, it's looking like that may be the way to save the dream.

Lee Cronin, a 40ish-year-old inorganic chemist at the University of Glasgow now "making matter come alive" in his lab with hefty research funding from the EU, expressed some interest in taking on the CERN challenge when we spoke recently. Cronin has also been in touch with Nordberg about this. However, not for a year, according to Nordberg. And there is, as yet, no updated list of participants from Cronin *et al.*, let alone a proposal.

As Nordberg explains in our interview later on in this chapter, the proposal does not have to be a lengthy document. The CERN-Atlas project continues to operate as a non-legal entity from a seven-page Memorandum of Understanding. And Atlas along with CMS, its sister experiment at the Large Hadron Collider, have produced evidence of the existence of Higgs boson! -- the "God particle."

Physicist Norman McCubbin says more about the CERN-Atlas MOU approach in his foreword to the book that Nordberg co-authored about the Atlas experiment -- *Collisions and Collaboration*: "Atlas and other similar experiments and enterprises are underpinned basically by trust, buttressed by collective self-interest and the 'currency' of mutual trust and esteem".

Lee Cronin did tell me one thing with certainty, that women will be part of any origin of life group he manages to put together in collaboration with CERN.

Women were left out of the 2011 and 2013 strategic meetings as presenters, seen as "not ready" to participate. And while I felt welcome as a journalist, I was later shocked and somewhat amused to hear that one European scientist attending advised his American colleague that I needed to be spanked.

Women are a crucial part of CERN's community of scientists. Indeed, physicist Fabiola Gianotti, the woman who led one of two teams in the discovery of Higgs boson, will now head CERN as of year-end 2015.

In retrospect, the exclusion of women as presenters at COOL EDGE 2013 was an omen. . .

My pre-COOL EDGE conference interview with Stu Kauffman follows.

October 29, 2012

"You're in the Village, you can't look at the biosphere," joked complexity pioneer Stu Kauffman, in a conversation from Seattle with me in Greenwich Village, just days before Hurricane Sandy shut down New York, leaving Manhattan south of 42nd Street powerless. As I now write by candlelight, another of Stu Kauffman's thoughts resonates: "Not only do we not know what will happen, we don't even know what CAN happen."

Moreover, systems chemist Günter von Kiedrowski, to whom Kauffman says he "handed-off" his brainchild -- his origin of life project at CERN -- after last year's initial workshop, further advised: "We can't travel back in time, we'll never know the historical course."

So exactly where does that leave the origin of life investigation?

Kauffman, von Kiedrowski, *et al.*, in their upcoming strategic meeting at CERN want to tap its phenomenal computer power to create chemically real reproducing and co-evolving protocells, and to apply some of CERN's organizational models to engineer an origin of life scholar grid that links experiments, researchers and sources of funding. Von Kiedrowski thinks the strong magnetic fields "the collider deals with" will be useful.

But protocells based on what definition of life? Presumably, this question will be high up on the list of topics of discussion in February 2013 at CERN. . . .

CERN-Atlas is hosting the Geneva meeting with John Templeton

106

Foundation covering travel expenses. Anticipated funds from the EU have not yet materialized but conference organizers say there has been significant dialogue and interest in underwriting research into the emergence of chemically reproducing synthetic cells capable of open-ended evolution. . . .

We were first introduced to Stu Kauffman's ideas in a big way in the 1980s at Santa Fe Institute, where he spent almost 20 years. In 2005, he moved to Canada and founded the Institute for Biocomplexity and Informatics at the University of Calgary, also serving as its director for five years. In 2009, Kauffman was a visiting professor at Harvard Divinity School. He's also been a visiting professor at the University of Vermont (2010-2012) as well as at Finland's Tampere University of Technology (2012).

Stuart Kauffman has a BA in philosophy, *summa cum laude*, from Dartmouth College, where he was Phi Beta Kappa and a BA in philosophy, psychology and physiology from Oxford University, where he was a Marshall Scholar. Kauffman began his career as a medical doctor, receiving his MD from the University of California, San Francisco.

Among Stuart Kauffman's many honors are: MacArthur Fellow; Doctor Honoris Causa, Catholic University of Louvain; and Fellow, Royal Society of Canada.

Kauffman has been issued a dozen patents (US, French, Dutch, Japanese) and is the author or co-author of five books.

Suzan Mazur: I understand that the origin of life meeting at CERN is your brainchild. Can you confirm who will attend? Are these some of the names? -- Markus Nordberg of CERN-Atlas and Mary Ann Meyers of The John Templeton Foundation, co-sponsors of the gathering; Eörs Szathmáry and Günter von Kiedrowski -- who are hands-on directing the project; plus Doron Lancet, Marcelo Gleiser . . .

Stu Kauffman: Yes, that's [some of] the unofficial list. . . Here's the story. My current position as Finland Distinguished Professor at Tampere University of Technology led, by sheer luck, through the "FEDIPRO" Board, to my meeting Markus Nordberg, who's of Finnish and Swedish descent and is co-head of the Atlas project at CERN. Markus said we (CERN - Atlas) want to use our computing power and organizational skills to help other fields. Do you know a field that might be ready to take off?

I said, "God yes, Markus, do you mean that? The origin of life field is getting ready to explode."

And Markus said, "Well, why don't you gather a group of about 25 or 30 people."

After years of working in the origin of life field, I have a bunch of friends, and so I called some of them and everybody said "Sure." About seven of us a year and a quarter ago had a brainstorming meeting on origin of life in Geneva at CERN. I have since handed off the origin of life project to Eörs Szathmáry and Günter von Kiedrowski, and they're fantastic!

Suzan Mazur: I saw the proceedings of the 2011 origin of life brainstorming workshop at CERN online. Are Dave Deamer and Harold Morowitz involved with the 2013 meeting?

Stu Kauffman: David and Harold are certainly part of the group I've pulled together.

Suzan Mazur: And Giuseppe Longo?

Stu Kauffman: Giuseppe is another wonderful story. Giuseppe is just a bloody amazing guy. He's a mathematician at the École Normale Supérieure, Paris. I hope that Giuseppe will come.

Suzan Mazur: I watched one of his YouTube presentations discussing Steve Gould's work.

Stu Kauffman: I have collaborated on something with Giuseppe I would love to tell you about. . . .

Suzan Mazur: Can we set down some of the basic reasons for your meeting at CERN?

Stu Kauffman: Yes . . . Here's what I know that Markus would say. He would say cautiously that CERN's business is physics, not biology. On the other hand, I am saying to him that somehow biology IS physics, it's just not particle physics.

So Markus would rightly say that where we are now is that CERN is committed, without a doubt, to help fund the meeting we're about to have in February. They have before them about nine proposals for computational experiments from the origin of life group that they've had for several months. . . .

Mary Ann Meyers has been wonderful. Thanks to her Templeton Foundation is funding the travel for participants and Markus is funding lodging and meals. We will organize the actual meeting of the 25 of us. There may be two meetings.

Suzan Mazur: And the funding to move forward is coming from Europe?

Stu Kauffman: Once again you need to speak to Günter. Aside from co-organizing the February meeting, Günter is a brilliant chemist. He's talking about making large scale instruments that are hundreds of NMR machines, for reasons that I understand partially. He was talking about this two years ago.

Günter, by the way invented systems chemistry. . . .

Suzan Mazur: I look forward to hearing more from Günter. Do you know what's on the agenda for the 2013 meeting?

Stu Kauffman: Yes. Certainly it involves collectively autocatalytic sets, whether they be peptide sets or RNA sets or DNA sets. It involves liposomes where Pier Luigi Luisi has shown that you can make liposomes that grow and bud and divide. Certainly it will involve getting liposomes to contain autocatalytic sets where Roberto Serra has shown theoretically that if they both divide, the two systems synchronize their division so that you've got a protocell. There's much to be added, but you've got the start. . . .

Now rather astonishingly, is something that I co-invented with Gábor Vattay and Samuli Niranen. A version may be realized in protocells. Gábor is a quantum physicist in Budapest and Samuli, a computer scientist in Finland. Two years ago we kind of discovered something we called the "Poised Realm," which Gábor thinks is a new state of matter which hovers reversibly between the quantum world and the classical world, back and forth. More and more we believe the Poised Realm is real and things can go from the quantum to being classical for all practical purposes and back. This is utterly novel, if true. And we don't yet know how this may apply to the origin of life.

I talked about Poised Realm when I met with the European systems chemists. And many of them liked it a lot. To my surprise, Günter and Eörs wanted it to be part of this meeting.

Markus tells me some of the CERN particle physicists are interested in the Poised Realm. I'm thrilled because I think maybe Gábor's right. Maybe we have discovered a new state of matter. If so, luck is upon us and it may have all kinds of valuable applications.

Suzan Mazur: Fascinating.

Stu Kauffman: For example, Suzan, it turns out the light harvesting molecules in cells show long-lived quantum coherence at ambient temperature. Physicists thought this impossible due to exponentially fast decoherence over a femtosecond. Quantum decoherence is thought to be one of the ways quantum systems become classical, at least for all practical purposes.

Decoherence is the loss of what's called phase information to their environment, if they're open quantum systems. Gábor has shown mathematically that Poised Realm has two axes (this is Gábor, me and Samuli), a Y axis goes from quantum to classical going up the Y axis and then to quantum going down the Y axis. Then there's an X axis, which you get by turning the Hamiltonians from order and criticality to chaos. Criticality is at the metal insulator transition for a quantum system.

And Gábor has done some calculations suggesting that criticality is particularly interesting where on the X axis many molecules seem to be located. We think it may have implications for drug discovery. But Gábor's the expert. He's done the real work.

110

Gábor has shown pretty much for sure theoretically that the rate of decoherence is very different at different points on the X axis.

If you're chaotic, decoherence can happen exponentially fast, which is what all the physicists thought. If you are either ordered or critical, or maybe just critical -- he's checking to be sure -- decoherence is very slow.

It's what's called a power law. So on Cartesian X Y axes, plotting logtime on the X axis, log coherence on the Y axis, you get a straight line down to the right. This is power law, very slow decoherence.

Well it turns out there are quantum effects that they found in biology in light harvesting molecules -- it's blown everybody away -- nobody knows how it can possibly be happening at room temperature because you're not supposed to get it due to exponentially fast decoherence. Gábor deduced why you should get this sort of power line decay.

A guy named Greg Engel in Chicago has shown that the light harvesting molecule does decohere in a power law way. So I think the Poised Realm may explain why there are quantum effects at room temperature.

Suzan Mazur: Thank you. What kind of life are you looking to create with the protocell at CERN? Are you using the classical definition of life -- chemical structure capable of doing Darwinian evolution? There are so many definitions now for what life is . . . .

Stu Kauffman: Well maybe and maybe not. Part of what we're going to do when we get together is decide what we want to investigate. . . .

[T]hey tried for 40 years to get single-stranded RNA molecules to replicate, perhaps hundreds of chemists, and they all failed. It should work. But it hasn't. And after 40 years or 50 years, you think -- maybe it's the wrong idea. People really tried hard.

You can't talk to Leslie Orgel, he died. But you can talk to Albert

Eschenmoser at the ETH in Zurich. He tried for years with DNA. He's about three years older than me, one of the best organic chemists in the world, and get his views about all of this.

Suzan Mazur: Do you know how the bench chemistry and computer chemistry ideally work in tandem to reach the goal, to get to the protocell? . . . .

Stu Kauffman: So the question is -- how did molecular reproduction ever start? Here is where theory may help, hence CERN computers may help. There's one theory for it. It happens to be mine. We can study experimentally the probability a given peptide catalyzes a given reaction then compute the peptide diversity expected to give rise to a collectively autocatalytic set experimentally.

And maybe we can come up with other theories about how such things could emerge spontaneously. Maybe it's not just about peptides and polymers. Maybe other things enter into it like temperature and energy, like Günter Wächtershäuser has talked about. Maybe it's clays like Cairns-Smith has talked about in systems chemistry. Maybe it's that lipids and liposomes orient molecules on their surface and the search problem for the molecules now is 2 dimensional not 3 dimensional. This changes the thermodynamics -- 2 molecules finding one another in 3D is much harder than in 2D, so you increase the rate of synthesis because things are confined to 2 dimensions. There will be all sorts of new ideas that will come forth.

Suzan Mazur: Why is such a project taking off in Europe and not the US, aside from the fact that CERN is in Geneva? Is there sort of a US - European rivalry regarding origin of life?

Stu Kauffman: There may be a US - European rivalry, Suzan. It's an historical accident. The accidents are the following. The US created the Astrobiology Institute. And the astrobiology program is wonderful. And it, to my understanding, is largely focused now on detecting life in the solar system, because it's part of the space agency. It's part of the NASA program.

Suzan Mazur. They're looking for universal life.

Stu Kauffman: Sure. Solar system. Exoplanets. Anywhere. But they seem not to be focused on creating life in a test tube here on Earth. I no longer know the astrobiology program. I knew it well for years. . . .

Suzan Mazur: You've said in a recent paper: "Evolution Beyond Newton, Darwin, and Entailing Law" that the biosphere arose "without natural selection" and continues to create its own future possibilities. In our conversation in 2008, you told me: "There are some physicists who are asking questions like is natural selection an expression of some more general process? Like entropy production. And it's all up in the air." Since then, Richard Lewontin has written in the *New York Review of Books* that Darwin intended natural selection as a metaphor, not to be taken literally. How do you currently view the issue of natural selection? Is it factored into or out of the making of your protocell?

Stu Kauffman: Thank you for asking and let me be as precise as I can. . . . No one ever asks the question within Darwinism and evolution: How did life start? Although Darwin famously said, what if there were a warm little pond with lots of chemicals in it, could life start in it? Somebody said well why aren't we seeing them now? Darwin said because somebody would eat it, some bacterium would eat it. . . .

If Giuseppe and I are right, there are no entailing laws for the becoming of the biosphere for the reasons we spell out. The phase space keeps changing in ways we cannot prestate. So we can neither write laws of motion for this evolution, nor, if we had them, could we integrate them. And if there are no entailing laws for the becoming of the biosphere, there cannot be a mechanism. Evolution is not a mechanism, it's certainly a process. Darwin was very careful to call it a process. So I agree with Darwin 150 years later, it's a process not a mechanism.

If Giuseppe and I are right, once we make protocells, they're going to start to evolve. And once they start to evolve, if what Giuseppe and I say is right, no law will entail how they evolve. I told this to Eörs and Günter, but the rest of the group don't know this yet. So we may be creating protolife and we can watch it evolve and we will not be able to say ahead of time what it will become. We will make something that is beyond Steven Weinberg's dreams of the final theory. . . .

But what Giuseppe, Maël Montévil and I wrote in "No Entailing Laws, But Enablement in the Evolution of the Biosphere Life" and posted on *Physicsarxiv* and in press, if true, changes everything.

[**Note**: Giuseppe Longo and Maël Montévil have since published the book: *Perspectives on Organisms: Biological Time, Symmetries and Singularities* with a foreword by Oxford University physiologist Denis Noble.]

Because if we could say, therefore, that the evolution of human life, culture and law and business and the economy is not entailed -- then reductionism, Weinberg's dream of a final theory, is false. Too bad. So think about the societal importance -- not only do we not know what will happen, we don't know what CAN happen. Then reason is an insufficient guide to living our lives and science cannot "know."

Flip a coin 10,000 times and calculate the probability that you get 5,223 heads. Can you do it? Sure, you can bring up the binomial theorem. Now that you know ahead of time all the possible outcomes of flipping the coin all heads, all tails, you know all the possibilities. And the statisticians say, you know the sample space. And if you know the sample space, you can construct the probability measure. But in the evolution of the biosphere, we cannot prestate the future evolution, which I call the "Adjacent Possible," so we do not know the sample space, so can construct no probability measure, so cannot reason about what we do not know "Can" happen.

If I'm right, once we make protocells, they're going to start to evolve. And once they start to evolve, if what Giuseppe and I say is right, no law will entail how they evolve. So we may be creating protolife that we can watch evolve but will not be able to say ahead of time what it will become. We will make something that is beyond the dreams of the final theory.

Suzan Mazur: What commercial applications do you envision if the origin of life pathway is found? Is it possible to think about applications at this point?

Stu Kauffman: The whole biosphere. You're in the Village, you can't look at the biosphere. Bu the entire biosphere was created by, co-created by evolving cells and multicellular organisms. So any way we can think to harness that could be of commercial interest. Let me point out a couple of things. In my book *Investigations*, I was trying to figure out what was an agent. You and I do things on purpose. We change the world to benefit it or ourselves if we can. So we're agents. I wondered what do you have to have to be an agent.

Here's what I came up with. You have to be a self-reproducing system doing a work cycle like a choo-choo train. If you do work cycles, you can vacuum the rug. You can build a mound of things. Gophers do. Bacteria do. So if we can make protocells that are evolvable, capable of doing work cycles, what can we do with that? It was enough to make a biosphere.

Can we do something in terms of drugs? I can't even begin to know. But rather than making nanorobots to do things, maybe we can make evolvable things to do particular clean-up jobs. For example, in systems biology a cancer cell has a unique contributing pattern of gene activity. What if you could make a drug or a protocell that recognized that specific pattern of gene activity of a cancer cell and kill it?

Another paper I wrote may find applications in protocells: "Answering Descartes Beyond Turing" -- at 72 you get to write grandiose titles like that. And this one certainly is.

In it is the idea of making an information processing system that's partially quantum and partially classical, so not an algorithmic Turing system, and imagining and realizing it in protocells. I think life does this. I think it is partially quantum, partially Poised Realm, and partially classical thus not describable by deterministic laws because single specific outcomes of quantum measurements are not determinate.

This is an entirely different non-algorithmic class of information processing systems, Trans-Turing Systems. We believe that they can be built. We have the funding from Lockheed Martin to do it and we're about to get €500,000 from the Finnish government to do it. My hope is that Trans-Turing Systems can be realized in evolving protocells.

Suzan Mazur: Does this have anything to do with CERN?

Stu Kauffman: CERN is on the edge of knowing about it because I or Gábor, preferably Gábor, is going to talk about it at the meeting in February.

Some particle physicists at CERN are very interested in finding out about this. It's only two years old, we invented it in 2010. And we'll find out.

Gábor is convincing me more and more that we weren't so stupid when we invented and discovered it. We're going way beyond the Turing Machine in our notion of information processing with the non-algorithmic Trans-Turing System, perhaps in a protocell created in collaboration with CERN.

Suzan Mazur: The origin of life computer grid that CERN may help you organize, will it be a democratic, open kind of accessible tool, for credentialed and non-credentialed researchers alike, including amateurs, students of all ages, etc.?

Stu Kauffman: Yes. It's just what you said. It's that we springboard the creation of an international web of stuff that's entirely new, not just origin of life, but the becoming of the biosphere and systems we create beyond entailing law. My hope is that it will be a wave of very new science and be open to everybody. Everybody.

In 1905, when Einstein was inventing relativity, do you think he could have gotten funding? He was a clerk in a patent office. Suppose he'd approached the funding grantors and asked what it was like to ride on a light beam, he would have been told, Albert stick with the patent office.

He changed the world. Radical science is hard to fund. So we need a way of funding science that brings together people of divergent views who respect one another, to listen in the chaos of not understanding one another, until they get to understand one another. So just the kind of a democratic system you're talking about is what we need.

Then you have collaborating teams and competing teams. And they collaborate and compete in such a way that ideas bubble up from the bottom and then get acted upon in a democratic way. This is important, beyond CERN. CERN is a model for reorganizing civilization. The reason is that at CERN you do not know ahead of time what you're looking for.

I mean, they were looking for the Higgs thing, but they didn't necessarily know how to go about it.

Suzan Mazur: Scientists are optimistic they'll have a protocell within the decade. But suppose computers fail to assist in the delivery of the secret of life.

Stu Kauffman: I'm going to say something really strange. If Giuseppe and I are right, computers can't get us there.

We're going to use the computers to try to help us do systems chemistry where we know what we're talking about by the way of tasks. Tasks are molecules, reactions and catalysis. . . .

In physics you can always prestate the phase space. That's one of Giuseppe's points. His other, and mine independently, is that the phase space of evolution keeps changing. Mine is that because of the argument I just gave you, you cannot prestate the way the phase space of evolution changes. So no law entails the evolution of the biosphere. And this means that you do not know what can happen, therefore, you cannot make a probability statement. Therefore, no algorithmic computer can simulate what will actually happen. It's beyond entailing law. The biosphere is more magical than that.

Suzan Mazur: I understand your point. . . . Are the concerns about the potential dangers of protocell technology justified or are the dangers being exaggerated?

Stu Kauffman: That's a completely legitimate question. Let me tell you past places where it's come up. It came up when we learned how to clone. And the NIH and the scientists doing it said: Stop, let's try to see if we can do any harm.

Well, we can do harm. We can evolve viruses that are more lethal. So no technology comes along that can't be misused, including the club. It's great for killing men two valleys over and stealing their women and kids.

So another area is random chemistry, where you make millions of random molecules and you ask: Can you make a bad molecule? Yes. The field has flourished. Can it be misused? You bet.

Now let's take your question about protocells. Do we know what they would do if we let them loose in the environment in Alabama? We have no idea. Therefore, do we have to figure out what it will mean to be careful with them if we're lucky enough to make them. Of course we do. But do we know yet? No.

The obvious thing to do first is a P-3 facility, so that if you've made them, they can't get out any more than bacteria could that could make something harmful.

Suzan Mazur: Making of the protocell is being projected to the public in a fearful way. Positive aspects are not being clearly articulated.

Stu Kauffman: We don't know what the possibilities are, Suzan, so that's why we can't.

Let me give you a historical example that you know. When the computer was invented, Tom Watson, Sr. at IBM said, there's going to be a market for three of these. And we're going to make the three. Well, he was wrong.

There are computers all over the world now and they've transformed it. Now we've got apps on everything. And we've got the World Wide Web and we're selling things over the World Wide Web and you've got Google and all these things emerged. And notice that nobody foresaw the Web and nobody saw us

118

selling things on the Web and nobody foresaw content on the Web that would create a niche in which Google could come and make a fortune. It's the same thing. We cannot say ahead of time what it is that we're going to co-create.

The world is more mysterious than we thought, yet we are co-creating it. Here are the possibilities -- the evolving economy. And let me show it to you in the econosphere.

Without anybody's intent, when Turing invented the Turing Machine, it led to the World Wide Web which led to Google. It was nobody's intent to create Google when the Turing Machine was invented. . . .

The computer created a niche for the personal computer created a niche for word processing created a niche for sharing files created a niche for the World Wide Web created a niche for selling things on the Web created a niche for Google created a niche for Facebook created the Arab Spring.

Suzan Mazur: Right, you can't predict.

Stu Kauffman: It means top down management and control is an illusion because you don't know what the relevant variables are going to be. Therefore, world government by the elites is nuts. They're trying to get us there, but it won't work. . . .

Günter von Kiedrowski:

Letter to an Origin of Life Circus Fan

GÜNTER VON KIEDROWSKI
(artist, Peter Sheesley)

In January 2013, Günter von Kiedrowski -- one of the world's great chemists, currently chair of bioorganic chemistry at Ruhr University Bochum, Germany -- took on the responsibility of co-organizing an origin of life strategic meeting of 25 scientists at CERN called COOL EDGE 2013. Von Kiedrowski is also considered by some the inventor of systems chemistry. Modestly, he says he only provided a suitable name for systems chemistry plus some basic ideas.

COOL EDGE was a private meeting that I was invited to cover. Von Kiedrowski sent the following email to me on January 17, 2013, concerning the conference. It is published with his permission.

Dear Suzan,

Thank you for letting me know about the Princeton event. I was an active member of the community of prebiotic chemists in ISSOL (the International Society for the Study of the Origin of Life) from 1986 - 2005 when I became extremely frustrated about this silly struggle between the "geneticists" (RNA-world believers), "metabolists" (iron-sulphur world/Wächtershäuser world), and "compartmentalists" embedded in NASA's astrobiology mission to gain control of the whole. So I didn't follow the invitation to China and I left ISSOL in the same year. I saw that the core idea of any of such theories was feedback, autocatalysis, crosscatalysis, and collectively closed autocatalytic sets and their coupling to the generation of information and the evolution of evolvability in the Darwinian sense. I also saw that the principle of feedback with respect to information transfer was poorly developed in chemistry. In chemistry, information is understood in its literal meaning (to inform means to give a thing form/shape) and thus information not only resides in constitutional information such as the sequence of biopolymers, but also in configuration (stereoselectivity) and even long lasting conformation (e.g., around chemical bonds with high rotation barriers). This is how systems chemistry came up. A mission towards a search for autocatalysis, cross.., collective.. in chemical systems based on the joint approach and cross-talk of prebiotic chemists, supramolecular chemists, and physicists, biologists, and computer scientists from complex systems research. This mission was successful: We have this crosstalk now (at least in Europe) and the pathway is towards integration of autocatalytic subsystems into "life like" supersystem.

Today I see, however, that even this picture is incomplete, as Darwinian evolution can be "directed" at the molecular level and so the question comes up whether it could become "directed" at later stages of evolution as well. Directed evolution always comes into existence when an "intelligent operator" defines the constraints for selection and thus "non-naturally" redefines what fitness is about. This has profound consequences even for evolutionary traits where the phenotype does not any longer map to a genetic layer in a direct manner. The evolution of human cultures is a matter of how fitness in "thoughtspace" is defined. There may be more than one thoughtspace. E.g., one in which fitness increases with "logical consistency"/objectivity and another one in which fitness increases with "emotional consistency"/subjectivity. The thoughtspace of egoism versus altruism: ME-culture versus WE-culture, and so on. We know that science, media, and finance are instances creating constraints for the evolution in thoughtspace/s and thus direct the way we live on this planet.

Close to my 60th birthday I have gained the impression that "Directed evolution" might be even a cosmic principle. I am not the first one who went into this "thought crisis". Exactly 40 years ago Francis Crick and Leslie Orgel went into such crisis when they published an article in Icarus (the astrobiology journal of these years), entitled "Directed panspermia". I don't know whether this crisis will be a "temporary distortion" or fluctuation in thoughtspace – or whether it will settle. It appears to me not bizarrely different from my previous worldview, just expanded. It plagues, however, my colleagues and my students -- for very understandable reasons. And I feel that my friends in the team of organizers are also a bit concerned about this attitude. Anyways, we will have two EDGE lectures in which we allow ourselves to become exposed to the unthinkable. We define EDGE as not yet accepted "protoscience" which might or might not become accepted in the future.

Please understand that this exposure of ourselves to the unthinkable must be handled with extreme care. We don't want to let the public draw the wrong conclusions, as you can imagine. While I personally feel grateful to you, and while we collectively think that the meeting IS important, we need to care about the consequences of our meeting. We don't want to damage the reputation of CERN. What might be arrangeable, but needs a new discussion of the four of us (CC) [Günter von Kiedrowski, Stu Kauffman, Markus Nordberg, Eörs Szathmáry], is a statement by the press that the EDGE lectures are a kind of "birthday present" to one of the organizers but that these lectures do NOT represent the opinion of CERN or any other scientist involved. We need to sort this out, Suzan, and I promise that we keep you informed timely.

With warm regards,

Günter

Show Preview

November 19, 2012

With the Large Hadron Collider safety issue now settled in the German court, the way is clear for a strategic meeting in February on origin of life at CERN. Some of the biggest names in science will be participating -- for starters, theoretical biologist Stuart Kauffman; Günter von Kiedrowski, considered the inventor of systems chemistry; Eörs Szathmáry, an expert on major transitions in evolution and one of the famed "Altenberg 16."

Markus Nordberg of CERN-Atlas is hosting the Geneva talks and John Templeton Foundation is sponsoring the gathering. Templeton supported the historic Vatican conference on evolution in 2009, and year after year helps underwrite the World Science Festival.

Stu Kauffman terms the event "radical science," aimed at finding an answer to the big question that interests every one of us -- who we are and where we came from.

But will the origin of life soufflé about to be baked in Geneva rise or fall?

Markus Nordberg, perhaps to avoid overexposure of the upcoming summit, has described CERN's role as simply one of helping to manage a "large multinational collaboration with a different kind of science."

But here's what some of the participants are saying about the meeting three months from now.

While there's a growing consensus among scientists that lipid-like entities may have been central to first life, Günter von Kiedrowski, a co-organizer of the event, told me, "we'll never know the historical course" because "we can't travel back in time."

Realization of that hard cold fact has led to a stepped-up interest in making a protocell ("a physical-chemical implementation of the simplest life form that either we can make or that can emerge spontaneously" -- Steen Rasmussen). So creating a protocell will be central to the Geneva talks, says von Kiedrowski.

Stu Kauffman, who says the concept of autocatalytic sets (*i.e.*, molecules catalyze each other's production) is his discovery -- although JBS Haldane may have been onto it first -- also claims the idea of an origin of life collaboration at CERN as his brainchild.

Kauffman says if all goes well, the group would like to tap CERN's phenomenal computer power to create chemically real reproducing and co-evolving protocells, as well as apply some of CERN's organizational models to engineer an origin of life scholar grid linking researchers, experiments and funding sources.

Many approaches to building a protocell are currently being explored. Roughly 10 to 100 or more labs, depending on who you talk to, are either mainly focused on synthetic cell development or on certain aspects of it. The approaches include not only the RNA (ribonucleic acid) world, PNA (peptide nucleic acid) world and autocatalytic sets, but cooperative feedback and computational protocells.

Some scientists, among them Danish physicist Steen Rasmussen -- who Stu Kauffman has told me he'd like to see at the February meeting in Geneva -- thinks researchers are now very close to having what is needed to synthesize a protocell. Rasmussen says this will "provide a brick in the ancient puzzle" of origin of life.

Rasmussen also says it is his deep fascination and awe for life that drives his research on the synthetic cell, as well as systems beyond. He asks: "How can we use the properties of living processes for the good of mankind, make technology that has some of these wonderful properties that living systems have?"

While Rasmussen -- considered by many to be the flag bearer of artificial life -- actually finds the idea of machines that can copy themselves and ultimately evolve a bit scary, he also says humans may be naïve to think that we are the end of evolution. But he doesn't think it has to be as frightening as Hollywood films project. Rasmussen envisions that we may "create things that can make more beautiful poetry, that can love more than we can."

Lab on a chip microfluidics is now unfolding, for example, which will enable at-home medical readings like cholesterol and PSA by simply plugging the test the chip has done into a computer port.

Rasmussen is also keen on John von Neumann's Universal Constructor, a mathematical machine that already exists, although it has not yet been implemented. "Our technological vision," he told me, "is that once we get to the point where we can combine the essence of biological systems as we develop with protocells bottom up at the microscale with what we do in factories, with what 3D printers do, that is top-down design, then we'll be able to have our own production facilities at home to print medicine, clothes, electronics."

But will the development of a protocell, take another decade, as many scientists predict?

"When I look back at what I said 10 years ago," Rasmussen says, "I did think we'd be able to do it in 10 years. We're not there quite yet. . . . We've put together an information-controlled metabolic production of the protocell components. We still haven't got evolution going yet."

Perhaps in jest, Freeman Dyson told me we should give it another hundred years. . .

But why are these origin of life strategic talks taking place in Europe and not the US?

Rasmussen spent 20 years in the US at Los Alamos and Santa Fe Institute, returning to Denmark five years ago to continue his research. He emailed me saying:

> "It was the shift in the availability of basic research funding that made me return to Europe. . . . Curiosity-driven research, also called basic research, has recently

grown in Europe and is more an integral part of the scientific culture there than in the US. . . ."

Kauffman says February's origin of life meeting will also identify other avenues of how molecular reproduction may have started. He expressed interest in Günter Wächtershäuser's ideas about temperature and energy and Cairns-Smith's regarding clays. Said Kauffman:

> "Maybe it's that lipids and liposomes orient molecules on their surfaces and the search problem for the molecules now is two dimensional not three dimensional. This changes the thermodynamics -- two molecules finding one another in 3D is much harder than in 2D, so you increase the rate of synthesis because things are confined to two dimensions."

Rasmussen describes the "wonderful and terrible" green fingers of chemists in relation to the origin of life, where one chemist synthesizes molecules without any problem and another using the exact same recipe "makes the soufflé fall flat."

Günter von Kiedrowski cautions that it's all "POP" (persistence of possibles) until all of a sudden and out of the blue it becomes real.

Kauffman concurs, saying we just can't predict the future, and that: "Not only do we not know what will happen, we don't even know what Can happen."

In other words, the proof in Geneva will be in the pouf.

Show Review

(*artist, Peter Sheesley*)

March 12, 2013

Covering a recent origin of life meeting at CERN on the outskirts of Geneva, I was surprised to learn that I'd been made a member of Atlas, the particle physics experiment at CERN. As such, I probably qualified for a room at the CERN hotel (dormitory), along with two dozen of some of the biggest names in science who had gathered for strategic talks end of February called COOL EDGE 2013. But as an independent journalist I'm used to making my own way, so I instead found lodging nearby at a centuries-old converted monastery housing a two-star Michelin restaurant midst vineyards overlooking the Rhone and bison in the field. Each day as I left the chateau for the tram ride to CERN where 3,000 scientists work to support the Collider, now under repair, I was in phase transition from the 15th to 21st centuries.

The meeting at CERN, invitation only, was hosted by the very savvy Markus Nordberg, a blonde Finnish-Swedish physicist turned Atlas administrator. It was designed, he said, for "friends of friends," *i.e.*, the metabolism-first crowd, who would likely keep bloody noses to a minimum (they did), with many of the talks touching on autocatalytic sets.

Said Nordberg sharing his strategic thinking in perfect British English:

> "When you invite friends of the friends, they don't argue too much . . . so that at least you have a little bit of critical mass . . . but then you have to expose it. That is to say, you organize a public meeting where the enemies of the friends are invited . . . it's polarized . . . people mix like hot and cold water. And that's the point at which you are able to set up a collaboration."

The larger focus of the four-day talks was to begin to discuss engagement with CERN as the brand, the facilitator and administrator of an origin of life "experiment." Ultimately, such an experiment has to somehow involve particle physics, which is CERN's main business. The proposal for a collaboration is now being cooked up.

Some of the scientists associated with the hoped-for experiment are drawn to CERN because of its computing power.

But Nordberg told me this about CERN's computers:

> "While we have computing power, this is not the only place. What makes CERN special is the way we handle it. We do it in a distributed way. . . [T]hese [origin of life] guys seem to want the opposite. They want to centralize power. Well, we don't have it. . . [W]e would have to reprogram all our computers. . . . Yes. It could be done. We could even be designing nice new motor cars. There's no problem. We could even design super jets. But that's not our mission. Why would we do it?"

The newer ideas presented in Salle Dirac hall at CERN came from Giuseppe Longo and Gábor Vattay. Longo is a mathematician at École Normale Supérieure in Paris and a collaborator of Stu

Kauffman, the complexity pioneer and "godfather" of COOL EDGE 2013. Longo, Kauffman and Maël Montévil share the view that there are no entailing laws in the evolution of the biosphere, meaning an end to reductionism. According to Kauffman, this means "reason is an insufficient guide to living our lives and science cannot "know"."

Gábor Vattay, a quantum physicist at Eötvös Loránd University in Budapest, also a Kauffman collaborator, discussed "poised realm," a state of matter he thinks hovers between the quantum and classical worlds. Lockheed Martin and the Finnish government have already seriously invested in the idea.

Vattay was largely applauded for his presentation with the caveat that he did not provide enough statistics. Some noticed that his low-slung jeans were also insufficient.

The big idea emerging from the meeting -- and it is estimated that it will take 10 years to actualize -- is to create an origin of life center at CERN.

Chemist Günter von Kiedrowski during the wrap-up session spoke in terms of a $1.5B investment to establish such an institute, proposing further that the building be named for Stu Kauffman with a Kauffman hologram adorning the front of the building.

Meanwhile, no one is even clear about how many scientists work in the origin of life field. Piet Hut, an astrophysicist at the Institute for Advanced Study, also affiliated with Tokyo Institute of Technology (Japan's MIT), told me at the meeting that he estimates there may be 500 or so origin of life researchers worldwide.

But how many independent and DIY (Do It Yourself) researchers are there? How many institutions are at work on origin of life? And what are the available resources?

While the COOL EDGE organizers -- Stu Kauffman, Günter von Kiedrowski and Eörs Szathmáry -- presided over daily sessions, the baton was later passed to two other scientists to further organize the group: John McCaskill, a theoretical biochemist at

Ruhr University Bochum, and University of Southern Denmark physicist Steen Rasmussen.

Both McCaskill and Rasmussen are actively running experiments that CERN "can understand," said Nordberg.

Some of the questions the origin of life group is considering follow, conveyed by John McCaskill:

> - Is origin of life experimentally reproducible?
> - Is origin of life process time and space compressible?
> - What is the nature of 2nd life?
> - How special are we?
> - How does origin of life influence the environment?
> - Can combinatorial physics accelerate it?
> - Did quantum mechanical coherence facilitate origin of life?
> - Experimentally what are the major transitions in origin of life?
> - What are the necessary conditions for open-ended evolution in life?

There was also some attention to "the unthinkable" -- a subject CERN distanced itself from -- during a 60th birthday party on February 27th for Günter von Kiedrowski, with his wife and students in attendance. Rupert Sheldrake was invited to speak about ghosts, but was unable to make it. Klaus Dona, a Viennese researcher of ancient art, replaced him.

I had a dinner engagement and left the lecture just as Dona was presenting the image of a giant footprint. Dona told me the Smithsonian has in storage bones of an unexplainable composition from a hugely tall human found among dinosaur bones.

Von Kiedrowski's mentor was British chemist Leslie Orgel, who toward the end of his life along with Nobelist Francis Crick wrote an article called "Directed Panspermia," that is, intentional, intelligent panspermia. Von Kiedrowski told me there is already theoretical evidence that dimensionality is a function of length of distance, referring to a recent lecture on the structure of space-time as a function of the landscape.

With Eörs Szathmáry, a commanding evolutionary biologist and one of the famed "Altenberg 16" scientists seated next to him nodding, von Kiedrowski told me that "it is very likely" that we are not alone.

Von Kiedrowski continued:

> "We have evolution on one side and this, of course, we cannot deny. But there is obviously another layer above evolution, which right now is in the layer of theological approaches because we cannot grab it. It doesn't allow us to do experiments with it. It is somehow separated from the real experience. So there is no difference between alien, extraterrestrial, inter-dimensional or deity. So if these things exist, then they will have technologies which surpass our understanding and we will have to give up the anthropocentric view as we had to give up the geocentric view."

The origin of life caravan moves on to Tokyo Tech's Earth-Life Science Institute, March 27-29, and an international symposium hosted by ELSI Director Kei Hirose. Hirose told me he too thinks "we are not alone."

Impresario Markus Nordberg

MARKUS NORDBERG
(photo by Claudia Marcelloni, courtesy M. Nordberg)

For all his savvy, it was clear Markus Nordberg had his hands full as CERN-Atlas host of the February 2013 origin of life conference called COOL EDGE. But somehow Nordberg never lost his composure during the week-long gathering of 25 supposedly like-minded scientists -- the metabolism-first crowd -- who now seem divided on how to move forward to pitch an origin of life project to CERN, as well as what to pitch to CERN.

Nordberg's official title at the time was Resources Coordinator of the Atlas project. Atlas ("A Toroidal Lhc ApparatuS") along with its sister experiment, CMS ("Compact Muon Solenoid") have now provided the world with evidence of the existence of Higgs boson, the so-called "God particle."

Nordberg arranged a charmed meeting at the end of the February talks, between the origin of life scientists and Sergio Bertolucci, CERN's Director of Research and Scientific Computing. But, again, there is now virtually no momentum.

Some of the scientists in the group have appeared disinterested because one of the requirements is for them to bring in funds to CERN to support their own project once they've got one. Others are turned off because the alpha males of the group dominated the February discussion. But Markus Nordberg remains optimistic that a way will be found to make an origin of life collaboration with CERN work.

Nordberg is a physicist with a PhD in economics. He currently heads Resources Development for Development and Innovation at CERN. Nordberg has been at CERN for over a quarter century. Some of his roles outside of CERN have included a faculty position at the University of Brussels and Visiting Senior Research Fellow at Centrum voor Bedrijfseconomie.

Markus Nordberg has been a member of the Strategic Management Society, the Academy of Management, and TUTKAS (the Association of Finnish Parliament Members and Scientists).

He is also author/editor (with Max Boisot, Saïd Yami and Bertrand Nicquevert) of the book *Collisions And Collaborations* about the Atlas experiment at CERN's Large Hadron Collider. The book looks at the organizational structure of CERN-Atlas, its teams of international scientists, how the project interacts with its suppliers, its contribution to e-science, etc.
In our interview that follows, Nordberg touches on CERN's influence in the wider culture as well.

February 28, 2013
Cafeteria, Building #40, CERN

Suzan Mazur: Can you tell me a bit about your background?

Markus Nordberg: I was born in the snow and grew up in the forest -- in the Helsinki region of Finland. I have two degrees, one in theoretical physics and another in business administration, from universities in Finland and The Netherlands.

Suzan Mazur: You've been here at CERN for 25 years.

Markus Nordberg: Yes.

Suzan Mazur: You came here right out of school?

Markus Nordberg: No, after completing my university studies, I came to CERN in 1989 for one year when the previous accelerator (LEP) started, just having completed my degree in physics. But I had then already decided I would go into business, which is why I wanted to do a PhD in economics. I was never planning to stay at CERN. However, I was intrigued by the question of why companies collaborate at a place like CERN. After all, it's very remote from market. It's very technical. So what's the benefit to society?

Suzan Mazur: You're moving on shortly to another position at CERN?

Markus Nordberg: Tomorrow.

Suzan Mazur: What will you be doing, can you say?

Markus Nordberg: It's about technology, funding and innovation. The idea is the scientific work that we are doing here, we're trying to make the results of our work more accessible to those in the industry in return for their collaboration. So instead of the old way of doing things where we would contract things, we would now ask them to participate in a collaborative mode. We would change the rules of the game with industry.

Suzan Mazur: The people you are collaborating with would help to bring funding for the project.

Markus Nordberg: Yes. We could try to bring in all sorts of funding like from the European Union.

Suzan Mazur: You're hoping that with the origin of life project, the individual scientists will have funding links.

Markus Nordberg: The origin of life initiative needs first to be defined as a project but yes, I hope the scientists will get funding.

Suzan Mazur: Was there an idea to make a grid, mapping out who the origin of life scientists are, the institutions, the resources? Was there a proposal to organize a grid using CERN's computational power?

Markus Nordberg: That's a good question. I believe the origin of life scientists are here to create that.

Suzan Mazur: Talking with Piet Hut the other day, he said there might be roughly 500 origin of life scientists worldwide. Who are they? Who are the amateur researchers? What are the institutions? How much money is being spent and who's doing the funding? Answers to those questions, I think, are necessary in moving forward.

Markus Nordberg: You can't ask me. I have to make clear that origin of life is not a CERN project. Particle physics is not biology. It's not chemistry. So why are we here? We are here because Stu, Günter, and Eörs are hoping CERN can help them with a brand name but also help the group get organized.

They are interested in the computing power. I have been telling them for two years that the right question has to be asked. We don't just give 100,000 processes to anybody randomly walking across the street, saying gee it would be fun to get a CERN process started -- because that process is delegated for particle physics, which we are paid for.

So if somebody comes from outside, we need to know three things: Who is it who is asking? What would they like us to do? Why CERN?

While we have computer power, this is not the only place. What makes CERN special is the way we handle it. We do it in a distributed way. Not everybody is doing it in a distributed way. In fact, these guys [the origin of life scientists] seem to want the opposite. They want to centralize computer power. Well, we don't have it. So we would have to reprogram all our computers. We'd have to do the reverse if we were to do what they want us to do. This is what we have understood from interactions we've had so far. But has a lot been really clear? This is one of the things I said yesterday.

Suzan Mazur: But could it be done?

Markus Nordberg: Yes. It could be done. We could even be designing nice new motor cars. There's no problem. We could even design super jets. But that's not our mission. Why would we do it?

Suzan Mazur: So no money would be coming from CERN, it would be coming through CERN.

Markus Nordberg: No money from CERN. You should speak with CERN management. My understanding and my belief is origin of life will never be a CERN project. It will never be an Atlas project. Unless the CERN council, which consists of 21 member states or 22 states, decides to change the mission of CERN.

Suzan Mazur: How can it work?

Markus Nordberg: CERN hosts, supports and -- I'll add -- facilitates experiments. Experiments are always a bottom-up thing. I've seen this hundreds of times.

Guys who like this get together. They may fight each other, but they're friends. They have brainy ideas. Some of them are probably nuts -- but let's be clear -- they're wonderful people.

You let them get the steam out. That may take years. But once the steam is out, they get to understand each other and they develop a common language -- which they haven't yet here. Then they are able to agree on a plan, and that plan they start making through a committee that CERN administrates. So the role of CERN in all of this is always administrative, facilitating, managing the process.

Suzan Mazur: Do you consider what the origin of life scientists are proposing to do in relation to CERN "an experiment"?

Markus Nordberg: I assume so. The guys first need to make a proposal of what they want to do. We can look at it and see what can be done, if at all anything.

Suzan Mazur: But you don't consider that a project?

Markus Nordberg: It's not yet a project. I'm talking about the CERN way, not specifically the way the origin of life group may see it. The way it works here is as follows: Physicists get together. They collaborate, design an experiment and send a proposal to committee, invited by CERN. If it is approved, it becomes a project.

CERN takes care of putting the committees together. But their members are not *de facto*. CERN folks, they are established physicists from all over the world. Usually people who don't have a vested interest in the particular project. Then if they think it is good enough, then they will recommend to CERN or to the CERN council or usually to the CERN management that this is a jolly good idea that should be supported.

Then the job of CERN, the director general, is to say, well now I have to fund this somehow. I have to put resources to it. But resources in terms of making sure that they have a hall somewhere where they can work together, that they have office space, that they can connect to email, that they have a place to have lunch. But the fundamental funding of the project itself comes from the collaboration. It does not come from CERN.

Suzan Mazur: So the scientists would have to bring in the money.

138

Markus Nordberg: Exactly.

Suzan Mazur: CERN is the brand name.

Markus Nordberg: Yes, providing infrastructure. The money can come through CERN to the projects, that's okay. Like we did for this workshop. The money came from the University of Vermont. It's not CERN money but we handle it for them. But people don't need to know that it came from Vermont. It's not relevant. In fact, I gave everybody a piece of paper -- you saw it -- and it has the CERN logo on it and it refers to Atlas. That's the only project I can put them in, as there's no OOL (origin of life) project.

Suzan Mazur: Are you disappointed that the meeting so far hasn't been more coherent in terms of producing a strategic plan?

Markus Nordberg: Yes and no. This is why these guys are here, to clarify their thinking.

Suzan Mazur: Did you expect there would be less noise?

Markus Nordberg: I'm actually surprised how little noise there is. There have been some very good points made but it's not that type of conference that would generate buzz.

Suzan Mazur: Wasn't the purpose of the conference to develop a strategic plan to go forward rather than nitpicking discussion about each scientist's model?

Markus Nordberg: I don't know if that's realistic at this stage. It would be nice if that happened.

Suzan Mazur: You think it will take a while for that strategic plan to materialize.

Markus Nordberg: My guess is that it will take 10 years.

Suzan Mazur: This group hasn't really come together as a whole before.

Markus Nordberg: That's exactly the point.   This is a first meeting where friends of the friends have been invited.   When you invite friends of the friends, they don't argue too much.   The ones who are not part of the club of the friends of the friends are not particularly happy about it.   There are two reactions.   They get publicly annoyed.   And then they criticize.

You can start with friends of the friends, so that at least you have a little bit of critical mass.   Some coherence in the group.   But then you have to expose it.   That is to say you organize a public meeting where the enemies of the friends are invited.   Anybody can come.

I mean, you just don't say you don't like a person because he thinks differently.   In fact, the way we do it is that if somebody says -- so Doron gave a wonderful speech, we would then hear from the number one critic of Doron who would then possibly say completely the opposite.   Now, of course, they usually do it the polite way.   We do it every time.   This is normal practice.   You never ever organize a public meeting where you are just giving one point of view.   Because that kills it.

What happens is that you have two sides and then you don't say let's have a vote.   But what that starts is a process polarizing the community because let's say there are some RNA versus metabolism views.   It's already polarized for some bizarre reason.

Suzan Mazur: It's coming together though.

Markus Nordberg: It seems to me it's polarized.   They will have to bring these guys together.   This is what takes time.   People mix like hot and cold water.   And that's the point at which you are able to set up a collaboration.   To be successful in getting money.   Because if they would now try to get money, the guys who are not in the boat are going to say "no way."

Suzan Mazur: Can you tell me what happened yesterday at the meeting with Sergio?   That he wants to see a proposal from the group.

Markus Nordberg: I think Sergio was a magnitude more helpful than I ever expected. I was extremely pleased. What he said was, you guys have to get organized. You have to be coherent in terms of what you want. He was urging them to set up an organization. This is what I mean by getting themselves organized.

Sergio sees the angle of the European Union. He's said he's willing to help once they've made out very clearly what it is they want to do. You can't just come to CERN. For instance, Suzan can't just come to CERN and say well I just had lunch with Sergio and I looked into his eyes and I repeated what Markus told me and he said yes, yes, yes. Yes, but it means nothing. It has to be a complete proposal, it has to be well thought out. What is it that you want? We don't know except for 100,000 processes.

These guys want to hear me telling them what it is they want. I said no. I would rather find a way of understanding what exactly it is they want. Today Doron made a very clear statement. They would like to identify specific patterns in all the simulations.

Suzan Mazur: I caught that.

Markus Nordberg: That just happens to be exactly the way. That I can relate to. I was very strongly making the point two days ago when John was talking about his experiments. That's very clear. It's tangible. You can measure it. The guy's already working on it.

And Steen, the Danish guy. These are people we can deal with because they are saying something that we understand. That's the first time in these discussions of two years that I've heard something I actually thought I understood -- at least on the conceptual level -- that I can do something with.

Suzan Mazur: Yours is a very important role. Scientists sometimes don't have administrative ability. What is the title of your new position?

Markus Nordberg: I can't discuss that at the moment, because we are still discussing it with the CERN management. This will take a couple of weeks. But as said, it has to do with the funding model which is what is special about the collaborations at CERN. Collaborations like Atlas, they are independent, but they are not legally independent. They don't have any legal status. Atlas is a non-legal entity.

Suzan Mazur: I see.

Markus Nordberg: Yes. But there has to be a mechanism that allows that to happen. So Atlas is a non-legal entity, described in a Memorandum of Understanding of seven pages without the annexes. If you read that Atlas document, you see that the management is elected every two years. There is a vote on it.

I've been doing this job now for 12 years. Last year I decided not to stand anymore for election. I just decided with my colleagues that now it was time to do something new.

And the news is – it was an observation that I've been making, which is very alarming. Because of the financial climate at the moment in public funding, fundamental technology development is hard to fund.

Governments are now -- I wouldn't say stalling, but they are very reluctant to make the funding available for the next 10 years. This is extremely dangerous. This is a personal comment. Some people wouldn't agree. But this is what I've seen. I can explain this in more detail why I believe this.

Suzan Mazur: I think you're right.

Markus Nordberg: You see the reason why this is important is that what is special about CERN is not so much the creation of technology. People think we invent new technologies. No. We further develop them. But it doesn't mean that the idea necessarily comes from us. This is a misconception. What we are good at is taking something and breaking it.

The joke that we always tell our suppliers is -- "If you make it, we break it." That's what we mean by creating it. By breaking it we make it better.

That's the idea of the collaboration. You take something from everybody, you break it, and then in that process you make it better. And then you give it back to the community.

For this reason, all collaboration members are expected to contribute intellectually. It is not a classical buyer-supplier relationship. For example, if a corporate lawyer walks in and says hey, we are ready to collaborate -- we'll give you 10 billion -- but we want all the patents. We'd say, thank you, but no thank you.

Because if we do that, we kill the seed of the innovation. That's the paradox. The paradox of innovation is -- if you really want to innovate and make a real difference, you can't protect it.

The best inventions ever are things that have been made public and the individual who did it got little, if anything, out of it -- which is considered in the US as dumb -- but in terms of the diffusion, it's usually the most successful way.

Tim Berners-Lee, is a good example of that, the inventor of the World Wide Web. Tim said I have used it, I hope it will be useful. I don't need to be thanked. I remember he was speaking very strongly about enabling access to everyone, people in Africa, etc. He's done it. That's great. And that's the spirit in which we do things at CERN.

Now, unfortunately, when you go to your funding agencies these days and you say I want to develop a chip and say it's going to be 10 times better than anything you can find on the market, maybe 100 times -- they don't usually buy that. They say no. You have to talk to, say IBM, or an industrial partner like that. You have to go talk to these guys. They want to see industrial collaboration, they want to count the money.

That's why we are changing our operating model. That we are now trying to use the collaborative model to work with companies. The idea of the model is that with collaboration we are able to tap into other resources. That is what I am now working on.

Suzan Mazur: Because there is not enough government funding available.

Markus Nordberg: Correct. We are looking at the European Union. They are not interested in the origin of life. They are not interested in the origin of the universe either. But they are interested in how technologies make our lives better. We can make a clear case that we are working together with scientists doing something extraordinary and we are bringing them in as collaborators. It's a philosophical change to the posture. Everybody talks about collaboration. But it's not equal -- we want to make it equal. So this is new. My next project is this one. I'm curious to see if it is workable.

Suzan Mazur: You're talking about, for example, CERN possibly being interested in developing ideas discussed at the meeting like giving life-like qualities to inanimate material.

Markus Nordberg: If you mean thinking about how science connects with society, yes. Our community has no other choice but to get people together, to collaborate on ideas. . . . We feel strongly that in order to make advances in our field, we have to get together in an organized way, in a new way. That's the only way.

The Smolin Factor?

GERRY F. OHRSTROM
(photo, courtesy G.F. Ohrstrom)

LEE SMOLIN
(photo by Nir Bareket, courtesy L. Smolin)

145

March 13, 2014

It was Ash Wednesday and the sidewalks were still in deep freeze outside The Standard, High Line Hotel in New York's Meat Packing District for the party following the premiere of Executive Producer Gerry Ohrstrom's *Particle Fever* at Film Forum. There were no celeb watchers, red carpets or black velvet ropes for this crowd of art-science enthusiasts. No parade of skeletal lovelies wearing Alexander McQueen and Dries van Noten. No flash of Louboutin red. This was serious, an event to celebrate the story about the search for Higgs boson, the so-called "God particle." And Gerry Ohrstrom wore his down vest.

Particle Fever easily captures the fever surrounding "first beam" at CERN's Large Hadron Collider, the "first collision" of protons, and the emergence of Higgs boson -- coaxed out of a vacuum by the Big Smash.

We see the optimism of Mike Lamont, beam operations group leader (plus his concern over setbacks):

> "The magnetic properties of the machine are good. The aperture is clear. There's nothing sticking into the beam pipe anywhere. These are very, very encouraging signs. Remarkable progress."

We see Peter Higgs in the CERN theater, the man who first proposed the particle decades ago, gently weep at the Higgs boson moment. Freeman Dyson looking on, all ears and smiling.

The Film Forum audience (both shows I saw were sold out) seemed intoxicated by things to come. But what might they be?

The promise says David Kaplan, a Johns Hopkins physicist and one of the film's producers in a voice reminiscent of Al Franken's, is:

> "I have no idea. We have no idea. . . What is the LHC good for? Could be nothing other than understanding everything."

Kaplan says further that they're doing it to reproduce in the LHC "the physics, the conditions just after the Big Bang."

Somehow the discovery brings us closer to a definition of life, which according to Harvard Origins of Life Initiative Director, Dimitar Sasselov, we so far don't have. The late Carl Woese agreed. In fact, he and Nigel Goldenfeld were awarded $8M by NASA in 2012 to investigate just that -- the principles of the origin and evolution of life.

Goldenfeld told me when we spoke at lunch in the garden of Santa Fe Institute last fall that he's clearly still in the throes of the investigation.

But Perimeter Institute theoretical physicist Lee Smolin, whose ideas seem to have ruffled some feathers at CERN, says he has one, and recently described life to me this way:

> "Life takes place in the context of a steady state far from equilibrium situation with a steady flow of energy driving it, and within that context it's a semi-isolated system surrounded and protected by a membrane by which it controls the flows of material and energy across in each direction in order to promulgate its own survival and reproduction."

I tried to get answers from several physicists at the party as to just why Smolin's new physics ideas were not embraced at CERN. But they would only go so far as to say that they were "not testable." I'd heard this before at the COOL EDGE 2013 Origin of Life conference at CERN.

Smolin counters that his ideas are testable and points to as an example his "cosmological natural selection," published in 1992 in which he says he made two falsifiable predictions, both confirmed by current data. And he describes the Perimeter Institute in Toronto, which he helped found 12 years ago, as the best in the world for theoretical physics.

In his latest book *Time Reborn*, Smolin notes that time is NOW, not an illusion, and that we live in a relational world:

"Embracing time means believing that reality consists only of what's real in each moment of time. This is a radical idea, for it denies any kind of timeless existence of truth -- whether in the realm of science, morality, mathematics or government. All those must be reconceptualized, to frame their truths within time."

So, if there was a fly in the *Particle Fever* soup, it was this. No alternative visions, such as Smolin's, were presented. Maybe there will be in a film sequel.

Particle Fever also highlights the new breed of women physicists with the camera frequently on Monica Dunford, at the time the film was shot a postdoc at CERN-Atlas.

We see Dunford's athleticism (and trophies) -- running, biking, rowing -- against the gorgeous backdrop of the French-Swiss countryside, plus her ability to command an audience at CERN. But we also hear her lament having to wear the hard hat and grunge. And somehow I was not surprised when she told me at the party following the film that she's now left CERN for a position at the University of Heidelberg.

CERN is still very much a man's world of steel-toed boots. Despite all the wonders of the LHC -- the largest machine ever built by human beings -- CERN lacks "powder rooms."

Chapter 6

Texas Wagon

Impresario Steve Benner

STEVEN A. BENNER
(photo, courtesy S.A. Benner)

December 6, 2013

Why exactly is the January 2014 Origins of Life/Gordon Research Conference in Galveston so locked down? There is an elaborate registration process, a fee of over a thousand dollars to attend, and approval by Chairman Steve Benner required. Plus, conference goers are asked not to reveal what went on there.

The trend is now for transparency. The January 2013 Princeton

University origins conference was streamed over the Internet. Even the COOL EDGE conference at CERN earlier this year invited media (me) without restriction.

I decided to have a chat with the chairman about the upcoming Texas origins meeting:

Suzan Mazur: You mentioned to me earlier this year at Princeton that as chairman of the January 2014 origins of life conference in Galveston, you wanted to invite Harry Lonsdale and the recipients of his 2012 Origins of Life research grants -- John Sutherland, Dave Deamer, Paul Higgs *et al.* You thought there would be results by the time of the Gordon Research Conference. I haven't seen any of those names on the program. Will they be presenting?

Steve Benner: I did invite them. I have not gotten a letter from Mr. Lonsdale. I did invite John Sutherland and a number of other people who are working in this area, and they had a schedule conflict. Something that was going on in the European Community. We've got an extremely strong representation from Japan.

Suzan Mazur: Japan is one of the sponsors of the conference.

Steve Benner: As you know, the Japanese government has just set up two major science and technology centers.

Suzan Mazur: I recently interviewed Kei Hirose, the director of ELSI (Earth-Life Science Institute), one of the centers.

Steve Benner: . . .We did, however, get Niles Lehman, who's at Portland State University.

Suzan Mazur: Niles Lehman -- Nilesh Vaiyda's collaborator on combining fragments of RNA. Vaiyda told me at Princeton that he and Lehman demonstrated that autocatalytic sets can emerge spontaneously.

Steve Benner: Lehman is going to have to represent the entire Lonsdale operation at this meeting.

[**Note**: Lehman in a separate conversation with me advised that at Gordon he will discuss the above experiment he did with Vaidya, not the Lonsdale-sponsored research that involves combining even smaller fragments of RNA.]

Suzan Mazur: Why has the approach to investigating origin of life now expanded from one of a sudden transition to one of a series of stages, evolution as a process before life materializes? It used to be that evolution was discussed once life emerged. I'm referring to the recent paper published by authors John Sutherland, Robert Pascal and Addy Pross, titled "Towards an Evolutionary Theory of the Origin of Life Based on Kinetics and Thermodynamics." The paper mentions Carl Woese in the second line -- Carl Woese, who told me in an interview last October, before he died, that LUCA, the Last Universal Common Ancestor, was not anything material -- it was a process.

Steve Benner: If you can send me a link to that paper, I promise to read it.

Suzan Mazur: Woese is cited. His former collaborator Nigel Goldenfeld is not. But Goldenfeld mentioned to me just weeks ago at the Santa Fe Institute that he agrees with Woese regarding LUCA. Sutherland *et al.* seem to be talking about the same thing. They're looking for process. LUCA is a process, nothing material.

Steve Benner: We have failed in any continuous way to provide a recipe that gets from the simple molecules that we know were present on early Earth to RNA. There is a discontinuous model which has many pieces, many of which have experimental support, but we're up against these three or four paradoxes, which you and I have talked about in the past.

The first paradox is the tendency of organic matter to devolve and to give tar. If you can avoid that, you can start to try to assemble things that are not tarry, but then you encounter the water problem, which is related to the fact that every interesting bond that you want to make is unstable, thermodynamically, with respect to water. If you can solve that problem, you have the problem of entropy, that any of the building blocks are going to be present in

151

a low concentration; therefore, to assemble a large number of those building blocks, you get a gene-like RNA -- 100 nucleotides long -- that fights entropy. And the fourth problem is that even if you can solve the entropy problem, you have a paradox that RNA enzymes, which are maybe catalytically active, are more likely to be active in the sense that destroys RNA rather than creates RNA.

Suzan Mazur: I think things are shifting to nonmaterial events.

Steve Benner: That's right. I think you're right about that. We have been trying for close to 10 years now to get what we call dynamic kinetic systems, a collection of small molecules interacting with each other, maybe some catalyzing transformations of others, a non-linear feedback, some kind of amplification and trying to find working examples, recipes, where you can actually go back and mix something and see something. We are finding all sorts of problems in getting behavior that we find useful, let alone Darwinian out of this. I'm hoping to walk out of the Gordon conference for sure with a clear understanding of how life originated by one of these schemes -- a dynamic scheme that involves A interacting with D interacting with C, back to A without my having--

Suzan Mazur: You note in the conference program that among the things that will be addressed is how big science might be funded in the future. You mention Hollywood and certain philanthropies as possible sources. Will you also be discussing how these funding sources could affect the research?

Steve Benner: The people at the conference will discuss what they feel like discussing. I have very little to say about it. What is somewhat new -- and Lonsdale reflects this -- the Simons Foundation has stepped up, the Templeton Foundation is interested. Of course, the National Science Foundation is actually talking about this. The Japanese government. The Max Planck Society is thinking about setting up a whole Max Planck institute on origins of life.

[Note: Max Planck Society could not find the right scientist to lead the research for its proposed origins of life institute and has now put its plans on hold.]

What is very much different for 2014, which I think you could really not say in 2012, when Lonsdale got started on this idea, is that many of the nontraditional private foundations are disappointed with the narrowly focused, health-related, hypothesis-driven-related research programs that avoid big science questions. The Gates foundation is maybe the exception. I'm amazed at how trivial the Gates Foundation calls are for little things in the corner of very narrow technology.

NASA, for many, many years, was the man. NASA was the organization that funded Woese continuously from the 1960s. A lot of the major science foundations out there in the world are stepping up and saying, "Yes, origins of life is maybe the kind of problem that maybe is now right. Maybe the chemistry, the biology, the physics is coming together."

Suzan Mazur: The public has been paying for scientific research but has not had a say in how funds are directed. Has there been any thought given to just reaching out to the public directly? That would assure the public that they're getting the science they want.

Steve Benner: I agree with that. We just had a discussion about crowdsourcing scientific funding. I've never seen it done for origins. Our group actually sells origins of life jewelry. The minerals that create the borate and the alkali and the phosphate are minerals like tourmaline and apatite and peridot.

Suzan Mazur: Are you going to be streaming the conference over the Internet?

Steve Benner: The Gordon Research Conference is defined by non-citability. You're not allowed to quote someone as having said something. Those are the rules.

Suzan Mazur: But the times are asking for more transparency.

Steve Benner: I know. I'll give you the email.

Chapter 7

Tokyo Wagon

Impresario Kei Hirose

KEI HIROSE
(photo, courtesy K. Hirose)

March 14, 2013

In January there was Princeton's publicly-streamed powwow on origin of life oriented to chemistry. Then February's private meeting on the subject at CERN on the Swiss-French border, focused on physics and theoretical biology. Tokyo's symposium, March 27-29, looks to the center of the Earth and beyond for answers to the origin and evolution of life.

Geophysicist Kei Hirose, director of Japan's Earth-Life Sciences Institute, is hosting the upcoming "1st ELSI International

Symposium" at Tokyo Institute of Technology (Japan's MIT).

Hirose is best known for identifying "post perovskite" in 2004, a mineral in the lowermost mantle of the Earth. Post-perovskite is a hexagonal crystal that can conduct electricity, and like mica, can be peeled.

Hirose fell in love with geology partly because it took him to exotic destinations. He admits he was a late bloomer academically, but he eventually caught up in a big way. In 2011, Hirose received the European Association of Geochemistry's Science Innovation Award as well as the Japan Academy Prize. A few months ago he was named director of ELSI.

It was long thought the Earth's core was 3.5 billion years old, but Hirose's research has led to a re-dating to roughly the time of the Cambrian Explosion 500 million years ago. This, he says, is one argument why it is important to continue to investigate the inner Earth in relation to the origin and evolution of life.

Kei Hirose's PhD is from the University of Tokyo. He was a postdoctoral fellow at Geophysical Laboratory, Carnegie Institution of Washington.

My recent interview with Kei Hirose follows.

Suzan Mazur: What is Japan's annual budget for origin of life research?

Kei Hirose: It's a difficult question. Origin of life is a big topic.

Suzan Mazur: I understand it's something like $80M over 10 years, but origin of life research is actually only part of that money. Is your Earth-Life Science Institute at the Tokyo Institute of Technology funded by the government or is it a private institute?

Kei Hirose: Our institute is funded by the Japanese government.

Suzan Mazur: What is the budget of your institute annually?

Kei Hirose: Almost $7M per year.

Suzan Mazur: What part of the overall program at your institute is origin of life?

Kei Hirose: $60M in total for 10 years for all the activities of the institute.

Suzan Mazur: How big a part of the institute is origin of life research?

Kei Hirose: Let's say half. Let me explain in more detail. The aim of our institute is to explore origin of life from the geological context. In order to understand the geological context, the geological environment at the beginning of the Earth, we have to understand the origin of Earth itself. Half of our principal investigators are geophysicists, geologists and astronomers. The institute itself has the grand aim of exploring origin of life, but half of our researchers are directly focused on origin of the Earth and its environment.

Suzan Mazur: But it's all related to origin of life.

Kei Hirose: Related but not all directly related.

Suzan Mazur: Are there other institutes in Japan working on origin of life aside from yours?

Kei Hirose: No, except for a small community organization. I don't know how many members it has.

Suzan Mazur: Do you favor the metabolism-first approach or RNA world-approach to origin of life?

Kei Hirose: Of course metabolism is very important, but it is not easy to understand metabolism for first life.

Suzan Mazur: I noticed you have invited to the 1st International Symposium at ELSI, March 27-29, key scientists whose research involves the RNA world -- like Irene Chen, Steve Benner and others who spoke recently at Princeton's origins conference. Do you favor the RNA world-approach or metabolism-first approach?

Kei Hirose: At this moment I do not have any preference. As

156

everybody knows, we need some breakthroughs to understand the origin of life. Here at ELSI we are considering each process from the point of view of the early Earth environment.

Suzan Mazur: Do you think the metabolism-first and RNA world approaches are coming together?

Kei Hirose: I don't have any preference at this moment.

Suzan Mazur: Why are you exploring deep-sea microbial ecosystems?

Kei Hirose: The deep sea hydrothermal system is one of the widely-accepted possibilities for origin of life. This is one of our points of focus. We have one principal investigator in charge of exploration of the deep-sea hydrothermal system. He is using submersibles. So we have access to those studies.

Suzan Mazur: Did you work with Bob Hazen at Carnegie Institution of Washington when you were there?

Kei Hirose: I spent a year and a half as a geophysicist at the Carnegie Institution. Bob Hazen was there at the time and I knew him, but no, we were not collaborators.

Suzan Mazur: I noticed that he was an invited speaker to your March 27 symposium, is he coming?

Kei Hirose: Yes, but sorry, I don't remember what he will be speaking about.

Suzan Mazur: You've also invited Tetsuya Yomo, who is known for his work on the minimal cell. Is he coming?

Kei Hirose: Tetsuya Yomo from Osaka University, yes, he is coming.

Suzan Mazur: Who are you working with at Harvard?

Kei Hirose: Jack Szostak.

Suzan Mazur: Is your view of origin of life algorithmic or nonalgorithmic?

Kei Hirose: I'm a geophysicist and so I'm not aware of each discussion on origin of life at this moment.

We have a team of specialists who are making a database of genomes in relation to the environment. Through computer simulations they relate the specific Earth environment to the specific genome set. If we provide the early Earth environment, then the computer simulations will tell us what kind of genomes there should be for first life.

Biologists will examine computer-simulated genome sets in specific experiments to verify the robustness of genome sets in the early Earth's environment. That's one of our processes.

Suzan Mazur: Are there philosophers presenting at your symposium?

Kei Hirose: No.

Suzan Mazur: Part of your institute's mission is to "critically examine the universality of these processes to determine the uniqueness of our planet with implications for the search for extraterrestrial life both in the solar system and beyond." Do you think we are alone in the universe?

Kei Hirose: That's a very fundamental question. No, I don't think so. We should have other life in the universe.

Suzan Mazur: Is there any evidence that you're aware of? Has the Japanese government shared any evidence, any cases, with the scientific community?

Kei Hirose: So far we have no evidence. But people are trying to find a way how to detect life in the universe. People are really thinking about what the critical conditions are to have life on the planet. We don't really understand at this moment what are the critical conditions for life.

For example, conventionally liquid water is critical for life. That means if you have oceans, then you have life. But it looks too simple. If you have very deep water, then we never have life. In this case you do not have nutrients for the first life. The nutrients, of course, include phosphorus.

It's not easy to find phosphorus in nature. Now we have enzymes which collect phosphorus from the water, from the ocean. But before we have enzymes, it's almost impossible to collect phosphorus from nature.

How to collect phosphorus from nature? Land is one of the strong possibilities. First life did not start in the deep ocean. Because if you have land, then there is weathering and that collects phosphorus from rock. Of course, we have water, but again, it should not be very deep ocean. So maybe the critical position is the presence of both land and ocean.

The oceans of the Earth now are just 0.02% of its weight, so oceans are a very tiny amount for the whole mass of the Earth. If you consider the materials originally on the Earth, they may include lots of water, 2% or possibly 5%. But we now only have 0.02% weight of water.

It's not easy to find lots of planets in the universe that have both oceans and land. So maybe we have very few planets with life. But I hope Earth is not the only one.

Part 3

Bottom-Up Tent

"Give it a hundred years, perhaps, but I don't think my prediction is worth anything. It all depends on what nature says, because nature is always surprising us. And probably in this case too. . . .

We are all aware of the dangers. That's nothing new. My feeling is there is no safe path. Trying to put a stop to biology is really not going to remove the dangers. Might as well let it go ahead. I'm quite aware of the risks in going ahead but the risks in not going ahead are probably just as big." -- **Freeman Dyson**, commenting to me about making life in the lab

Chapter 8

The Metabolists

Freeman Dyson

FREEMAN DYSON
(photo by Randall Hagadorn, courtesy F. Dyson)

"Then after that came us -- stage six. That's the end of the Darwinian era, when cultural evolution replaces biological evolution as the main driving force. "Cultural" means that the big changes in living conditions are driven by humans spreading their technology and their ways of making a living, by learning from one another rather than breeding." **-- Freeman Dyson**, Eastover Farm

June 27, 2012

Several weeks ago the Lonsdale prize went to researchers who think first life was RNA, a replicating creature. So I rang up Freeman Dyson, emeritus professor at the Institute for Advanced Study, Princeton, to see if he still embraces the idea that original life was a "garbage bag world," a membranous creature with dirty water trapped inside that reproduced for a billion years or more with high rates of error before replicating.

Dyson does the math on this in his book, *Origins of Life* (2nd. ed.), based on Alexander Oparin's cell-first theory of metabolism. Dyson calls it his "toy boat model," calculated with pencil and paper, where thousands of molecular units make the leap from disorder to order with "reasonable probability."

During our phone conversation, Dyson told me that he does indeed still hold to his hypothesis and also still thinks RNA was a byproduct of that first creature's own metabolism, emerging as a parasite and eventual symbiotic partner. He says it doesn't make sense that original life copied itself without getting its act together first.

In his *Origins* book, Dyson also refers to Doron Lancet's work on defining metabolism, also based on Oparin's model, by computer simulations of origin of life, saying:

> "Doron Lancet has tackled this problem by studying computer models of the evolution of molecular populations, which he calls replicative-homeostatic early assemblies (RHEA). In these models, metabolism is defined in a general way as the evolution of a population in which some of the molecules catalyze the synthesis of others. He finds conditions under which populations can evolve to a high and self-sustaining level of catalytic organization."

This prompted me to call Doron Lancet, a professor at Weizmann Institute, to see what his current thinking is. I reached Lancet at a conference in Stockholm. He had this to say about Freeman

162

Dyson:

> "He was my first inspiration. A chapter in his book, *Infinite in All Directions*, made me realize in the early 1990s that DNA/RNA was not necessarily the holy grail, and that there was an alternative in the form of molecular assemblies composed of mutually interacting simple molecules. The "Lipid World" model as a viable alternative to the "RNA World" would not have come to be without him."

Dyson envisions seven stages of life: (1) garbage bag stage -- metabolism without replication; (2) parasite emerges from garbage bag creature, parasite can replicate but not metabolize – "zooms around" cells as viruses in packages; (3) collaboration of metabolic and parasitic creatures, RNA invention of "most mysterious" ribosome; (4) coupling of metabolic and RNA worlds (modern cell), two billion years of speciation and sex, beginning of Darwinian era; (5) multicellular organisms; (6) "us" and end of Darwinian era with cultural evolution replacing biological evolution; (7) whatever happens next.

Freeman Dyson describes himself professionally as a mathematician. He is recognized as one of the architects of quantum electrodynamics as well as one of the most provocative public intellectuals. As a scientist he has not needed a PhD, his BA in math from Cambridge has been more than enough. He now holds 21 honorary degrees.

Dyson's ability to find music in problems may reflect on his father having been a composer. As the story goes, as a teenager he, young Freeman, dealt with the challenge of reading Vinogradov's *An introduction to the theory of numbers* by learning Russian and then translated Vinogradov's text into English.

Freeman Dyson is a Fellow of the Royal Society, National Academy of Sciences, and Paris Academy of Sciences, and the recipient of numerous awards, among them: Lorentz Medal (Royal Netherlands Academy of Sciences); Hughes Medal (Royal Society); Max Planck Medal (German Physical Society); Enrico

Fermi Award (US Department of Energy); Lewis Thomas Prize (Scientist as Poet, 1996); Templeton Prize (2000).

Dyson's books include: *Disturbing the Universe*; *Weapons and Hope*; *Origins of Life*; *Infinite in all Directions*; *From Eros to Gaia*; *Imagined Worlds*; *The Sun, the Genome and the Internet*; *A Many Colored Glass: Reflections of Life in the Universe*.

Aside from his affiliation with the Institute for Advanced Study, he serves on the board of trustees and is a past president of the Space Studies Institute, founded by the late Gerard O'Neill. Dyson is also a member of Jason, a defense advisory group of scientists.

My interview with Freeman Dyson follows.

Suzan Mazur: Has your thinking about origin of life changed significantly since the second edition of your book *Origins of Life* 13 years ago? And do you still say everyone is equally ignorant about origin of life?

Freeman Dyson: Yes, I would still say that. We've learned a certain amount since then but the basic mystery remains. We don't have a clearly defined chemical path from a mixture of garbage to an organized cell. That's what we somehow have to discover. The main difference between my thinking about things then and now is that now we firmly understand that there was an RNA world. The evidence for the preponderance of RNA at some stage has become stronger. And, of course, we know about various kinds of RNA operating in living cells today.

Suzan Mazur: Is your hypothesis still that the original living creatures were cells with a metabolic apparatus but without a genetic apparatus -- this was not an RNA world -- and that cells reproduced with high rates of error for perhaps at least a billion or more years before they began replicating? And that the replicator was a parasite. You say in your book that by metabolism you mean "what the Germans mean by *stoffwechsel* with no restriction to genetically directed processes."

Freeman Dyson: Yes, I would still say that there must have been

164

a purely metabolic phase. My thinking is that there were three origins: the original metabolic life, followed by parasitic RNA life, and then finally protein life. There is evidence now that RNA, the replicating stage, probably started a little earlier.

Suzan Mazur: In your book you have a section on Oparin's theory of cells first, enzymes second and genes third. And of Doron Lancet's work, a proponent of Oparin's theory. Do you continue to be keen on Lancet's "Lipid World" regarding origin of life? And if so, why?

Freeman Dyson: Yes, although I don't know what Doron's been doing recently.

Suzan Mazur: You describe Lancet's idea in your book by saying, "Life began with little bags, the precursors of cells, enclosing small volumes of dirty water containing miscellaneous garbage." Can you say more?

Freeman Dyson: Yes, that certainly I would agree with. Certainly lipids are the most likely candidates. It would be the first stage of developing life. These little bags which gradually become cells. I would say that still looks very plausible. But I don't know what further research Doron has done in the last 10 years.

[**Note**: Doron Lancet has emailed me saying:

> "Freeman Dyson's mention of my work relates to our basic idea that life began with much smaller and simpler molecules than RNA . . . Dyson was attracted to this idea (which he jokingly, but lovingly called the garbage bag model) because it constituted a quantitative, physicochemically rigorous model for self-replicating entities that could jump-start life without resorting to long templating biopolymers such as RNA . . .
>
> What is so nice about such spontaneously accreting building blocks is that (1) they can be simpler than nucleotides, the building blocks of RNA, *e.g.*, fatty acids (simple lipids) found in meteorites as per Dave Deamer's papers and (2) they come in thousands of different kinds,

allowing the assembly to be "interesting" compositionally and chemically, hence jump-start more complex subsequent chemistries, including the formation of simple polymer chains, as claimed by both Dyson and Deamer [Deamer was on sabbatical in Lancet's lab 10 years ago]. . .

In short, we claim (as Dyson and Deamer do) that RNA is a result, a consequence of simpler chemistry, not a necessary early "moving force." What is good about our model is that it is not hand-waving. We simulate in the computer accurate chemical progressions that mimic the process described for the appearance of RNA-like templating molecules."]

Suzan Mazur: What do you think of the Lonsdale origin of life prize going to RNA world proponents John Sutherland and Matthew Powner? The search was for a proposal that identified life as "a self-sustaining chemical system capable of undergoing Darwinian evolution."

Freeman Dyson: I think I would not make any comment about that. I don't know anything about the details.

[**Note**: Doron Lancet says further:

"RNA World and Lipid World are considered by many as competing models for life's origin. John Sutherland partially alleviated some of the most bitter criticisms against the RNA World by showing that the small molecular building blocks of the RNA chain may be formed prebiotically. This is still far from saying it wins as a full-fledged model for the origin of life. . . . There are many open questions that continue to be debated in the field regarding formation of chains, the capacity of such chains to replicate and more."]

Suzan Mazur: So you still think it was metabolic and cell first.

Freeman Dyson: Yes, the RNA world was clearly a stage, but I would call that stage two, not stage one.

166

Suzan Mazur: You say the time frame may be a little sooner for the RNA world. How much sooner would you say?

Freeman Dyson: Completely unknown. We have no idea really how fast things went.

Suzan Mazur: How many camps would you say there are regarding the origin of life question?

Freeman Dyson: I haven't really been following all the discussions.

Suzan Mazur: Are you familiar with the ideas of Adrian Bejan that origin of life is 100% physics and that both animate and inanimate are live systems, *i.e.*, organized flows of matter and energy?

Freeman Dyson: I would strongly disagree. It's essentially a problem of chemistry.

Suzan Mazur: Who else's origin of life work on this subject do you like?

Freeman Dyson: I haven't really been following it, so it's hard to answer. But it obviously is a problem of chemistry, and the chemists have not made much progress as far as I know.

Suzan Mazur: What kind of experiments should they be doing?

Freeman Dyson: Little tiny microdroplets experiments. Chemistry now is getting the tools to do that. Getting accustomed to working with nanograms of material. You want to have experiments with big numbers of little droplets with various mixtures of stuff and seeing what happens

Suzan Mazur: You draw an analogy in your book between origin of life and the origin of body plans half a billion years ago, a "sudden efflorescence of elaborate body plans," during the Cambrian explosion. Have you had further thoughts about this in light of the "evo-devo revolution"? Did form come first or did form arise from genetic programs?

Freeman Dyson: I haven't thought about that recently. Clearly we are understanding much more about embryonic development. . . .

Suzan Mazur: Do you think form came first?

Freeman Dyson: By the time of the Cambrian explosion is very late in the history of life and genetics had become very powerful. But, of course, we have no idea what happened in detail.

Suzan Mazur: How soon do you think we'll get to the bottom of things regarding origin of life, *i.e.*, make the breakthrough?

Freeman Dyson: Give it a hundred years, perhaps, but I don't think my prediction is worth anything. It all depends on what nature says, because nature is always surprising us. And probably in this case too.

Suzan Mazur: A hundred years. You think it's going to take that long?

Freeman Dyson: Well, I would call that short.

Suzan Mazur: You've been criticized for making blasé comments regarding genetic engineering, in light of what we now know about cell dynamics and jumps. Would you comment?

Freeman Dyson: We are all aware of the dangers. That's nothing new. My feeling is there is no safe path. Trying to put a stop to biology is really not going to remove the dangers. Might as well let it go ahead. I'm quite aware of the risks in going ahead but the risks in not going ahead are probably just as big.

Suzan Mazur: You've opposed recent wars. Do you ever have regrets about your comments about the atomic bomb and Hiroshima or of having served as a military adviser?

Freeman Dyson: Oh not at all. I still am a military adviser. I'm very happy with that. The military needs people from the outside to give them advice and to bring them into contact with the outside world.

Suzan Mazur: You're obviously advising them against wars.

Freeman Dyson: They don't ask me when they're starting a war, of course. I only advise them about details. We don't expect to be asked whether or not to start a war.

Suzan Mazur: You now consider yourself a natural scientist?

Freeman Dyson: I'm a mathematician basically. What I do is look around for problems where I can find useful applications for mathematics. All I do really is the math and other people have the ideas.

Suzan Mazur: Would you like to comment further about origin of life?

Freeman Dyson: I don't think so. I've been interested just now in the Prisoner's Dilemma, which is a totally different problem that has to do with the evolution of altruism. A very important problem, but nothing to do with origin of life. I just published a piece in the National Academy of Sciences about the Prisoner's Dilemma.

Suzan Mazur: I notice in your book that you have a section on Lynn Margulis, who you describe as one of the "illustrious predecessors," particularly for her work on symbiosis, saying that "she set the style" in which you came to think about early evolution. Do you miss her?

Freeman Dyson: Yes. Oh very much, yes.

There's another lady who I admire very much -- Ursula Goodenough. Do you know her?

Suzan Mazur: I don't.

Freeman Dyson: She's also a biologist who does very beautiful things. She's an expert on speciation, the way new species are developed. Anyway I recommend her, she's just about as good as Lynn Margulis.

Suzan Mazur: That's quite an endorsement. . . .

169

Doron Lancet: The Origin and Synthesis of Life

DORON LANCET
(photo, courtesy D. Lancet)

"One of the greatest challenges of Synthetic Biology is to chemically engineer a simplified self-reproducing system. Even the simplest of such systems, obtained by a top-down approach, *i.e.*, by gradually simplifying present-day living cells appears to be irrevocably complex. Therefore, a bottom-up approach is more appropriate if Synthetic Biology's declared goal of combining science and engineering in synthesizing novel biological functions and systems is to be followed. Self-reproduction comes in two different flavors: one is the ability of individual molecules, such as DNA and RNA to generate their own copies. Indeed, many Synthetic Biology efforts rely on this unique attribute, attempting to develop and utilise molecular replicators such as a self-copying ribozyme. The other reproduction flavor is the capacity of entire cells to fission and form progeny. In our laboratory we take the latter route. . . . Because present-day test-tube biochemistry is rather limited, we resort to computer simulations. . . . This approach echoes new developments in the realm of systems chemistry, a joint effort of prebiotic and supramolecular chemistry, as well as theoretical biology and complex systems research to address problems relating to the origin and synthesis of life." -- **Doron Lancet**

August 8, 2012

Since the investigation into who we are and where we came from increasingly involves creating a living cell with modern chemicals that may not have existed on early Earth -- should we even continue to call the research origin of life? Weizmann Institute professor Doron Lancet says this dichotomy of origin of life and synthetic life or artificial life is itself artificial, and that scientists exploring origin of life are indeed working in artificial life science as well, including him. Moreover, Lancet does not see much difference between origin of life science building a living cell from the bottom-up and synthetic biology working from top-down (think Craig Venter) -- meaning there is plenty of controversy ahead since scientists are predicting a protocell will likely become reality within the decade. . .

Doron Lancet has distinguished himself from many of his colleagues in the field by proposing that life on Earth could have started with any mix of chemicals and reproduced, and that it did not have roots in a replicating world of polynucleotides. His model -- metabolism came first -- does not rely on RNA and DNA.

Lancet's ideas for this approach to life's origin were referred to as the "garbage bag world" by Freeman Dyson in his book *Origins of Life* a dozen years ago. Dyson still shares Lancet's perspective on this, as well as Alexander Oparin's 1920s cell-first thinking, and says life may have reproduced this way for over a billion years or more before the appearance of RNA.

Over the last 15 years, Doron Lancet has been doing the math on these concepts as part of his "Lipid World" investigations using a computer system he calls GARD (Graded Autocatalysis

Replication Domain).

In 2010, a *PNAS* (*Proceedings of the National Academy of Sciences*) paper by Eörs Szathmáry (of "the Altenberg 16") and colleagues challenged Lancet's GARD model saying it did not exhibit evolution because compound genomes lost properties.

However, Lancet *et al.* counter that after reexamining the Szathmáry analysis and focusing on clusters of composomes "appreciable selection response was observed for a large portion of the networks simulated."

Lancet has more recently noted in an email to me:

> "The paper by Szathmáry *et al.* has examined a very special (and to a large degree untenable) case of our model, which casts a shadow on the critique. I should at the same time say, that like many other simulated models of evolution, GARD is far from being ideal, and will require long research to make it show what we call "open ended evolution." We are working exactly on this point. . . ."

Lancet says he spends his mornings investigating genetic diseases, afternoons thinking about "the quiet little pond" of origin of life research, and evenings speaking to public audiences about science.

Doron Lancet received his BSc in Chemistry and Physics from Hebrew University of Jerusalem and his PhD in Chemical Immunology at Israel's Weizmann Institute of Science. He did postdoctoral work at Yale and Harvard.

Lancet is currently Ralph D. and Lois R. Silver Professor of Human Genomics at the Weizmann Institute in Rehovot, where he also teaches bioinformatics. He is director of Israel's National Knowledge Centre for Genomics and of the Crown Human Genome Center at Weizmann. Lancet also serves as president of ILASOL (Israel Society for Astrobiology and the Study of the Origin of Life). He is a member of the editorial board of *Biology Direct* and the advisory board of the Lifeboat Foundation as well as a member of HUGO (Human Genome Organisation) and EMBO (European Molecular Biology Organisation).

His pioneering of olfactory signals led to his receiving the R.H. Wright Award in Olfactory Research in 1998. Other awards include the Hestrin Prize from the Israel Biochemical Society (1986) and the Takasago Award of the American Association for Chemoreception Sciences (1986). Lancet is the author of more than 180 scientific papers and several patents.

He has written science columns for *Ha'aretz*, as well, and he and his Lancet Group developed the web-based encyclopedia on human genes called GeneCards.

Writing with Barak Shenhav, Doron Lancet contributed to the 2009 MIT text *Protocells* with a chapter called, "Compositional Lipid Protocells: Reproduction without Polynucleotides."

I spoke recently by phone with Doron Lancet in Israel. Our conversation follows.

Suzan Mazur: Would you bring us up to date on where you and your colleagues are now in testing your idea that life began in a "garbage bag world," as Freeman Dyson described it, of self-assembled metabolic creatures that possibly reproduced for a billion or so years without relying on RNA?

Also, since no one knows exactly what the conditions were on Earth wherever and whenever life emerged, and we may never know, is yours an investigation into original life or an exercise in creating artificial life?

Doron Lancet: Let's begin by talking about conditions for life's emergence. The specific conditions matter less. There should be some provisos, or general conditions without which I don't see how life could have begun. For example, if people tell you life began at a temperature of 25 degrees Centigrade, 110 degrees or 360 degrees (in suboceanic vents) -- this doesn't matter. What does matter is the principle of what would constitute acceptable molecular roots of life, and at the same time have sufficient simplicity to warrant emergence from an abiotic mixture of chemicals.

I would also state the following: We don't know, and we will likely never know what were actually the exact chemical substances that began life. But we can wisely guess what principles such chemicals had to obey.

One of the worst mistakes, in my opinion, and in the opinion of Freeman Dyson and to some degree also of David Deamer, is that people think of first life in terms of life as we know it today -- with 20 amino acids and four nucleotides, the latter being a rather elaborate molecule with a nitrogen base, a sugar and a phosphate. And they say: "That's how life should be. That's how life should have been from its inception." That's utterly wrong, in my view.

Suzan Mazur: Thank you. That's important to establish.

Doron Lancet: Let's discuss my ideas, then think about their testing. Dyson wisely called this general class of ideas the "garbage bag world." What Dyson meant by garbage bag world was the principle that any set of molecules could have jump-started life. That set of molecules need not have been polypeptides or amino acids or nucleotides or polynucleotides (such as RNA and DNA) or for that matter fats, lipids, sugars or any other of the known array of molecules entertained as those that began life.

A chapter in Dyson's book, *Origins of Life* discusses the ideas of Alexander Oparin, the Russian biochemist who in the 1920s wrote a book that marked the beginning of modern thought about the origin of life. This is where Dyson writes about "garbage bag world," relating to Oparin's model of "coacervates," tiny chemical spheres that fuse, split and copy themselves. In that context, Dyson describes a random collection of molecules in a bag that may contain catalysts causing the synthesis of other molecules that in turn act as catalyst to synthesize yet other molecules, thus allowing "garbage bag" copying.

I'd like to clarify the basic mechanism involved: Molecules, when left alone, may go in different chemical directions undergoing conversion from A to B to C to Z to Q, etc. If there is a catalyst around, it will direct reactions, making them more efficient and focused. So the idea is that when you have a bag of molecules, irrespective of what they are, the constituents will begin to interact and help each other form additional similes, paving the route to "bag replication." Attempting to provide evidence for this kept us busy for nearly two decades.

To quote Dyson:

> "[T]he bag may be growing by accretion of fresh garbage from the outside, and the bag may occasionally be broken into two bags, when it is thrown around by turbulent motion. The critical question is then: what is the probability that a daughter bag produced from the splitting of the first bag with a self-reproducing population of molecules will itself contain a self-reproducing population?"

I share with Dyson the notion that such chain of events, while of low probability, is a likely path from dead chemicals to life, as indicated by our computer simulations.

Importantly, there may be billions of different bags of molecules formed in this ancient ocean, in the primordial soup, and very few of the bags will have the capacity to produce a simile of the entire bag when its molecules interact. These precious few unique bags with specific sets of molecules will feebly reproduce and may jump-start selection and evolution. . . .

Suzan Mazur: How are you testing this idea currently?

Doron Lancet: How do you test black holes and galaxies? Often by mathematical formulations and computer simulations, to do what you can hardly do in a laboratory, as is the case for the multi-million years of life's origin.

Suzan Mazur: You're largely relying on computer chemistry rather than bench chemistry.

Doron Lancet: Yes. Mainly. I say mainly because there are experiments we and others do to prove aspects of our theory.

Suzan Mazur: Do you consider your investigation one into origin of life or of creating artificial life?

Doron Lancet: Both. It's more origin of life because this question really challenges us. And it is very far from being solved.

Suzan Mazur: Would you say the origin of life community is largely focused on discovering how first life arose or creating artificial life, a protocell?

Doron Lancet: These two worlds interact with each other. . . . This dichotomy is artificial. Every scientist who works on the grandiose question of how life began sees how answering some of the questions addressed above will be useful in constructing artificial life. Many of us are interested in both aspects. We are interested in the big question that is far from being solved, and we're interested in the down-to-earth agenda items related to potential mechanisms that can lead to artificial life. It is very difficult to get funding for the grandiose question, so a lot of us get our funding for topics related to synthetic or artificial life.

Suzan Mazur: You're saying it's easier to get funding for synthetic life research than origin of life research?

Doron Lancet: The simple short answer may be yes. But we have to define what artificial life is. Artificial life typically is taking part of existing life and making very simple systems in which these existing parts play important and relevant "games." This may also be a valid route to understanding primordial events.

Bottom-up artificial or synthetic life uses materials (*e.g.*, protein and DNA) from present day life, as does top-down synthetic life.

But there is another kind of bottom-up research. Our own 15-year old model called GARD (Graded Autocatalysis Replication Domain) is a concrete chemical system of fat-like substances (lipids) whose behavior is emulated by a set of non-linear equations. When run in a computer, such systems exhibit the capacity to replicate, to undergo mutations and selection and rudimentary evolution. GARD's focus is on molecules not necessarily present in life now. As in the garbage bag world, our

176

"Lipid World" molecules are contained in "bags" and exert catalysis on each other, to enact "bag reproduction."

Using the principles governing the interaction of these molecules, we are investigating whether this can lead to a simile of life, entities with sufficient chemical complexity to be called "primitive life." The interactions within such systems can be described by linear or non-linear mathematical equations alike.

Admittedly, the molecular bags described by our GARD model start out very simply, and to show a valid path of scrutiny, need to become more and more accurate, with a higher and higher specificity and fidelity. The question is: What is the simplest conceivable collection of molecules endowed with these capacities that show rudiments of selection?

Suzan Mazur: And this selection that you're talking about is artificial selection, since you're dealing with an artificial situation. Right?

Doron Lancet: When you describe something with an equation and simulate it in a computer, it is not necessarily less natural than nature itself. You can simulate the formation of a planet from a nebula. You can simulate the formation of the moon from the Earth. The atomic bomb has been designed to explode according to physical-mathematical equations. Are these equations "artificial"? The same may apply to equations that describe selection, be it in a population of present-day insects or in early protocells.

Notably, without a computer, galactic dynamics or protocell dynamics may encompass a million years, which we cannot wait for. Computer simulations may often be the only way! . . . .

Suzan Mazur: What is the point of inventing artificial, synthetic life?

Doron Lancet: I thought we were going to discuss origin of life?

Suzan Mazur: You said you're looking at both.

Doron Lancet: The reason for investigating artificial life is double. One of the best ways to understand life is laboratory generation of simpler life. And artificial life can also help in understanding the origins of life. That's why I'm interested.

Suzan Mazur: But how might synthetic life revolutionize medicine, industry and information science? Isn't that why there's an investigation, for applications and commercial opportunities?

Doron Lancet: Artificial life and synthetic life -- these are very, very broad domains. At one extreme are applications potentially important to medicine, how to cure disease, the circuitry of gene expression, etc. Understanding how life began on early Earth is at the other extreme. In between is a whole graded spectrum. Whether rung 30 of the ladder is more important than rung 40 is not what's significant about the story.

What is wonderful about the story is that in the last century we've been able to understand life as a complex chemical system that we can make simpler and simpler without losing the properties or attributes of living cells.

The simplest living cell we know has 500 genes with molecular machines that convert DNA to RNA (RNA polymerase) and molecular machines that can read the RNA and synthesize proteins accordingly (ribosomes). It has membrane all around with many chemicals dissipated in it.

A protocell has yet to be made, but we can assume it will have very little of all of that, because the ribosome is such an utterly complex machine. There is no way it could be contained within an honest-to-god early protocell.

On the other hand, there are many other scientists, wonderful scientists, productive scientists who continue to take bags made of fat, lipid vesicles or liposomes and stick inside them a ribosome and an RNA molecule and some building blocks for protein and watch what happens. It cannot have been how life began, because the ribosome is too complex to have been there early on. Still it's important to look at this kind of protocell.

Suzan Mazur: Is what David Deamer is doing by mixing in the mononucleotides to come up with a replicating cell along the lines of what you envision regarding origin of life?

Doron Lancet: If you are asking could mononucleotides have existed inside a bag of fat on early Earth? -- yes. But, if you ask could the mononucleotides form orderly chains, including polynucleotides such as RNA or DNA? -- it's more complex, because you need specific catalysts for that. But you can think of a way by adding a few other smart molecules that would then interact with each other to form short chains of polynucleotides.

Suzan Mazur: Is the public sufficiently informed about synthetic life?

Doron Lancet: The public is utterly uninformed about science in general, including the serious attempts taking place to understand how life began. So let's try to make them informed.

Suzan Mazur: What should some of the societal concerns be regarding artificial life?

Doron Lancet: I see no problems. The artificial life that we discuss is so far from being able to do anything wrong or good that it's not an issue.

Suzan Mazur: But scientists are saying that within 10 years they're going to have a protocell. And, according to *Artificial Life* editor Mark Bedau, although the immediate danger is low, these protocells could in the future "be able to metabolize material from their environments, reproduce, and evolve" and "could cause problems for human health or the environment" if control is lost.

Doron Lancet: Yes, there will be some entities that will be capable of making copies of themselves in a very simple way but this is not an issue. You can now kill people very easily with chemicals you synthesize in the lab, it doesn't have to be a protocell. Natural viruses kill people as much as designed viruses.

When it comes to protocells that we're trying to cook up in the laboratory, either by computer simulations or by experiments, these are very, very far from being dangerous.

Suzan Mazur: Do you think a certain percentage of research budgets regarding synthetic life or even origin of life should be allocated to raise public awareness?

Doron Lancet: For science education, so people can read and share in the joy of scientific discovery.

Suzan Mazur: What about oversight of synthetic life? Watch dog organizations.

Doron Lancet: Thirty years ago genetic engineering scientists were confined to laboratories that had 16 doors so nothing got out. It then went down to 10 doors, and then to 5. Now those doors have been continuously open for years and nothing happens. Artificial life is much less dangerous than recombinant DNA. Recombinant DNA speaks the same chemical language that controls our own cells. Artificial life, typically, speaks a foreign language and does not easily interact with cells in our human body, but I admit there would be exceptions.

Suzan Mazur: Do you think we'll see a protocell within 10 years? You've previously commented as follows:

> "It is obvious that a primitive protocell in the form of an assembly of arbitrary organic molecules cannot be expected to undergo reproduction-like dynamic changes at rates that are measurable in standard devices."

And you've said that "it could take months, years, or even decades," meaning handing off the experiment to the next generation.

Doron Lancet: I was talking about a specific type of primitive protocell that, as mentioned above, uses a different "language" than present-day cells. Also, science is erratic and hard to predict.

We promised to cure cancer in 10 years 50 years ago, though we are much closer to finding a cure. And life is changing dramatically because of modern biomedicine and computer technology. The humble effort we are making in our laboratories in terms of artificial life is not likely to have such immediate impact on our lives. But in the long term it will. In 10 years there will be first glimpses. I don't doubt that in 10 years something extremely important could happen, but this may still be very far from affecting our lives as much as biomedicine and computer technology.

What is nice about origin of life science is that it's a quiet little pond in the middle of the forest, away from the hustle bustle of the city of modern biomedical science. There people can contemplate the interface between chemistry and biology and how life could have emerged as well as how to create it. I don't think we have to worry about some kind of artificial life monster jumping out threatening human survival in New York, Tel Aviv or anywhere else.

Chapter 9

The Compartmentalists -- Michael Russell, Elbert Branscomb, Nick Lane: "Oomph" & Origin of Life at Hydrothermal Vents

MICHAEL RUSSELL
(*photo, courtesy World Science Festival*)

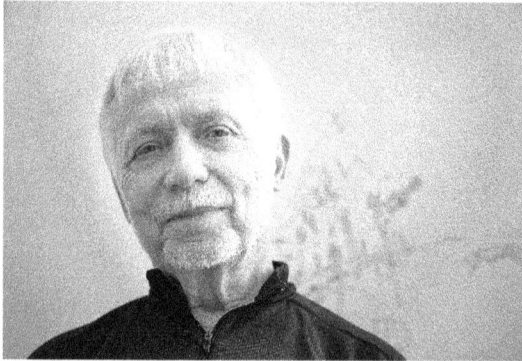

ELBERT BRANSCOMB
(*photo by Kathryn Coulter, courtesy IGB, UI-UC*)

NICK LANE
(*photo, Wide-Eyed Entertainment, courtesy N. Lane*)

"Life is an incredible thousand-ring circus of twirlers and high-wire trapeze artists, a fury of nonequilibrium activity. It requires power to drive it. Our contention is it had to be that way right from the start when the circus was vastly simpler." -- **Elbert Branscomb**

August 6, 2014

The notion that life originated in hydrothermal vents was for a long time a sleepy area of scientific inquiry because the vents first found, known as "black smokers," were way too hot and acidic. But in 1989, Michael Russell, a British geochemist who now heads the Planetary, Chemistry and Astrobiology Group at NASA's Jet Propulsion Lab - Caltech, proposed the idea that there were also alkaline hydrothermal springs. He began exploring serpentinization -- a geochemical process where seawater circulating through oceanic crust reacts with minerals found there to produce hydrogen, etc., and warm alkaline fluids. Russell suspected that, in effect, serpentinization set the stage for life to emerge in Hadean times when the oceans were still anoxic and highly acidic. He envisioned that a Hadean ocean was also home to a protected environment in the form of porous, precipitous mounds produced by serpentinization. These hollow castles or submarine alkaline vents -- with their complex walls of microcompartments and micro-channels separated by film-like mineral membranes -- Russell believed, could have served as a "hatchery" of first life with the right energetics.

Interest in Russell's now 25-year old alkaline springs hypothesis began to peak in 2000 with the discovery of Lost City, an actual alkaline hydrothermal vent system on top of the Atlantis Massif in the Atlantic Ocean. So far the Lost City system is unique.

The formation resembles hot wax paraffin plunged into a tub of cold water, or mini-volcanoes. But Lost City is not a volcanic system and its internal fluid temperature is 50 - 90 degrees C instead of 400 degrees C, as in black smokers. Marine geologist Deborah Kelley, who led the first expedition to Lost City, describes its fluid as "similar to liquid drain-o" (extremophiles can live in it).

If, in fact, Russell's ancient submarine hatchery of life once existed, what could have been the bioenergetics, since ultraviolet radiation was unlikely to be the necessary spark?

Data collected from Lost City plus lab simulations by Russell and his collaborators, Wolfgang Nitschke (Centre National de la Recherche Scientifique (CNRS)), Marseille), Elbert Branscomb (University of Illinois, Urbana-Champaign) *et al.* has led them to think that the energetics central to the emergence of life in the Hadean hatchery would not only have been the catalysis provided by metals, but also the power "conversion" magic due to proton flow involving the vent system's mineral membranes being exposed to alkaline conditions on one side and acidic seawater on the other.

The idea is that this movement of protons then got "baked in" to the emerging life system. The system remembered how to move protons to create fuel, which in today's life is ATP (adenosine triphosphate) -- all cells burn it to keep us alive. Russell, Branscomb *et al.* have elaborated on these points in recent articles published in *Philosophical Transactions of the Royal Society B* and *Astrobiology Journal*.

Indeed, says University College London biochemist Nick Lane, UV is the last thing life emerging at hydrothermal vents would have wanted.

The thinking is that alkaline springs life would at first have been looking to stay close to home, *i.e.*, in the hatchery, where the water was cool rather than venture out and about into an inhospitable world beyond the vent.

Lane recently co-authored a piece in *Science* on life's origin at hydrothermal vents with Bill Martin (University of Dusseldorf) -- who argues contrary to Russell, Branscomb and Nitschke that "metal ions alone" could have provided the necessary energetics to get life going.

Following are excerpts from my interviews on origin of life at hydrothermal springs with Michael Russell and Elbert Branscomb peppered with comments from Nick Lane. Bill Martin was on vacation at the time I spoke with Russell, Branscomb and Lane, and could not be reached for further comment. I'll be brief with introductions, since these are well known personalities in science with lots to say.

Michael Russell pioneered the field of alkaline hydrothermal vents a quarter century ago. He is a British geochemist with a PhD from the University of Durham and has for the last half dozen years been investigating alkaline springs theory at NASA's Jet Propulsion Lab -- Caltech. Prior to that Russell was a research fellow at the Scottish Universities Environmental Research Center. He's also spent time prospecting for precious metals, among other great adventures, including studying volcanoes in the Solomon Islands.

Elbert Branscomb is a physicist at the University of Illinois, Urbana-Champaign, formerly chief scientist of the US DOE Genome Program, and before that a physicist at Lawrence Livermore National Laboratory. He was awarded the Edward Teller Fellowship in 2001 for his contribution to LLNL and DOE.

Branscomb has a PhD in theoretical physics from Syracuse University.

Nick Lane is a biochemist at University College London where he investigates origin of bioenergetics and origin of life. He is also a masterful book author (*Power, Sex, Suicide*: *Mitochondria and the Meaning of Life*, and other provocative titles) -- winning the Royal Society Prize for Science Books (2010), symposium organizer, and plays Irish fiddle too, among other talents. Nick Lane earned his PhD from the University of London.

Suzan Mazur: Are we talking about the origin of life or the origin of bioenergetics?

Elbert Branscomb: One of the deep and controversial issues is in what sense are those the same question.

Suzan Mazur: You're saying the origin of life and origin of bioenergetics are one and the same.

Elbert Branscomb: To an important degree, yes. **Life is an incredible thousand-ring circus of twirlers and high-wire trapeze artists, a fury of nonequilibrium activity. It requires power to drive it. Our contention is it had to be that way right from the start when the circus was vastly simpler.** That the first step in life's origin was stumbling onto a solution to initiate power, overcoming an energetics challenge. The information side of life -- RNA, DNA, protein synthesis, polypeptides and the rest -- that came later, it's derivative.

The critical issues are: What were the exploitable sources of power the inanimate world supplied, and how were those sources of power tapped to drive the first high-wire chemical acts of the circus of life?

Several of the chemical activities needed for the most rudimentary metabolism, the most rudimentary biochemistries are ones that can not happen on their own. They have to be powered, driven against their thermodynamic will. Forced.

We think we have an idea where the power could have come from and how the hook-up happened.

Suzan Mazur: There are some substantial scientists with varied perspectives investigating origin of life and bioenergetics at hydrothermal vents.

Elbert Branscomb: Yes. No important scientific question has ever been resolved by a conflict-free community walk in the park of objective "evidence." For example, there was that famous 30-year struggle in the field of bioenergetics known as the "ox-phos" wars over Peter Mitchell's idea of the proton motive force and the process he called chemiosmosis.

Proton motive force is a membrane-spanning difference in proton concentration -- a "gradient," or "disequilibrium" -- which living cells constantly drain and replenish using the power obtained by "burning" food.

It was Mitchell who realized that this gradient was used by life to drive the production of its main intracellular chemical fuel "ATP" -- which it uses in turn to power most of its intracellular circus. The proton gradient acts like water behind a dam. The disequilibrium is used to drive molecular generators as the protons flow downhill through a membrane. The generators are literally turbine-like rotary motors, molecular motors. They charge up the ATP chemical fuel system.

This was one of the most important discoveries made in the science of life, but battles over the idea persisted even after Mitchell got the Nobel Prize in 1978. Some of his opponents had to die for the fighting to come to an end.

Oddly, whereas Mitchell's idea was right on the principle, it was in some basic respects wrong on the mechanism. It was Paul Boyer who fundamentally got that part right for which he was awarded the Nobel Prize in 1997.

Suzan Mazur: Michael, you trailblazed the serpentinization idea. Why do you think the earliest bioenergetics was proton gradients via serpentinization at alkaline hydrothermal vents rather than metal ions alone?

For example, Filipa Sousa, and Bill Martin -- a former collaborator of yours -- at the University of Dusseldorf, reported the following in a recent Elsevier paper:

> "First newer findings document eyebrow-raising similarities between the bioenergetics reactions of anaerobic autotrophs and geochemical reactions that occur spontaneously at some types of hydrothermal vents, an exciting development. Second, electron bifurcation has recently been discovered, a mechanism of energy conservation that explains how it is possible for acetogens and methanogens to reduce CO_2 with electrons from H_2, even though the first segment of the reaction sequence is energetically uphill."

Michael Russell: The reason why Elbert, Wolfgang and I think the proton motive force is important and have written recent articles -- one published by the Royal Society and the other a story for *Astrobiology Journal* -- is because it's so peculiar that life would pump protons out of a cell just to have them return to generate an energy currency within the cell. It's rather like pumping water uphill just so you can use the free energy on the way downhill in a turbine to give you electricity. Why bother to pump them outside or up the gradient in the first place?

So we think it's because the conditions in which life started offered just this gradient for free. Life is stuck with this rather peculiar way of working. That is, microbial cells evidently have to pump protons out of the cell to get into the periplasm because those were the conditions at the very origin of life in one of these hydrothermal springs where the protons were already on the outside. So that was one point.

It's naïve for Sousa and Martin to say that these are new ideas. The idea is about 20 years old. We said this a long time ago. When they say geochemical, what they're missing is how do they get the methyl group? The CH3 group. All life has CH3, many CH3 groups amassed in its organic molecules. Our view is that it can't have been produced from the reduction of carbon dioxide because the thermodynamic barrier is too high. And I notice now that they don't suggest that either. Amazingly. But they suggest a geochemical origin after all. In their *Biochimica et Biophysica Acta* paper [cited above, published by Elsevier] they say that they are gaining their methyl group from hydrothermal means.

Suzan Mazur: The one I just quoted from where they say, "it is possible for acetogens and methanogens to reduce CO2 with electrons from H2, even though the first segment of the reaction is energetically uphill."

Michael Russell: We agree with that, but the reduction only gets you carbon monoxide. It only takes away one oxygen from CO2. And I think they're saying the same thing. It only took one oxygen out of carbon dioxide to make carbon monoxide.

We suggested using electron bifurcation in 2009. Now suddenly Bill has adopted the idea, well established in the literature -- that electron bifurcation was significant, but I think we were first to say how significant it would be to the origin of life. We involved molybdenum as a redox bifurcating element in our 2009 paper. And then Bill wrote quite an extensive paper about electron bifurcation. So he kind of got into the same field as us.

What they're saying is they need a methyl group just like we do. If you look in the *BBA* paper, they will admit, in parentheses, that Tom McCollom says no methyl sulfides have been found in hydrothermal springs. And yet they say, well, there's methane there so there must have been some methyl stuff.

Suzan Mazur: What about the point Bill Martin, Filipa Sousa and Nick Lane make in the recent *Science* magazine article, "Energy at life's origin" about "the synthesis of high energy bonds that underpin substrate level phosphorylation can be catalyzed by metal ions alone--

Michael Russell: I have no sympathy with that view at all. You can't "catalyze" the synthesis of "high energy bonds." Such syntheses have to be driven.

Suzan Mazur: --and does not require either proteins or membranes whereas chemiosmotic synthesis of ATP requires both. This indicates that substrate phosphorylation came first in early bioenergetics evolution and powered the evolution of genes and proteins."

Michael Russell: There are just so many things laid out on the table here. Let me just address the one point about the methyl group. What they're saying is that they need methyl groups from geochemistry. That's their big thing. They've absolutely got to have it, and yet in the *BBA* paper in parentheses Tom McCollom says they've never been found. I presume Tom McCollom was a referee and that's why they had to put that in. I'm guessing that. So, in other words, they have no geochemical source for the methyl group, the CH3 group.

Nick Lane: I think we all agree that methyl sulfides could be important, whether they are delivered from the deep by serpentinization as CH3SH -- I don't know. Bill wants CH3SH out of the ground; there's not much evidence that it is available, but it would be surprising if it wasn't; Mike wants CH4 out of the ground, and we know that this does indeed emerge today from Lost City, but we can't be totally sure it is abiotic.

I want H2 out of the ground, some HS- and would like to react them all at the interface in the vent to form CH3SH and CO directly, driven by the proton gradient across thin semi-conducting inorganic barriers.

[Bill Martin has returned from vacation and responded:

190

Yes, Lost City effluent is low in methylated compounds, but it contains even less CO2 than it contains methylated compounds (if we exclude methane, a "methylated" compound of which there is a lot). The water in those systems comes from modern sea water, which contains quite a bit of CO2. Why does the CO2 disappear? Where does all that CO2 go? It is consumed, before the water reaches the surface, by very hungry microbes that live in those systems. For microbes, methylated compounds are an even juicier morsel than CO2, they are consumed too, also by microbes. So Mike's concerns about methyl groups is well-taken, but there are abundant microbes at these vents today that consume all the good stuff like CO2 and reduced intermediated en route to methane. Methane is left over because it is so stable and so unwilling to react, it has very high bond energies, much like teflon. Before life's origin, there were no microbes within vents to consume CO2 or methylated compounds. So Mike's comparison raises an interesting point (thanks Mike!). And if Mike has a suggestion for methane synthesis that does not entail methyl moieties, I'm all ear!]

Suzan Mazur: Elbert has told me this:

"The catalysis that Bill Martin is invoking is absolutely critical and a huge thing, but it couldn't provide the kind of coupling of outside sources of energy to the inside thermodynamic lift that I think is required."

Elbert Branscomb: What's wrong with resting the case on catalysis, the ability of metal ions, particularly in the precipitous mounds that are produced by the serpentinization mechanism -- the thing that is wrong in saying that's sufficient is that all that a catalyst can do is accelerate the rate of the reaction which would go on its own. To get life started, I think it is inarguable, and I think Lane would support that. Lane has argued that strongly himself many times -- that to get life started you have to do more than just make chemistry go that would run on its own. . . .

The wonderful thing about the Russell model is that it provides the proton gradient driver for free via geochemistry. Early life, therefore, did not have to do what we now all do -- which is to get our mitochondria to recreate that proton gradient at great difficulty and expense to drive ATP.

Nick Lane: What Bill's specifically saying, and the details do matter here, and that's why my name is on this [*Science*] piece -- is that chemiosmotic coupling with proteins across a membrane came later, across an organic membrane.

Suzan Mazur: When Bill Martin talks about "metal ions alone" and the earliest bioenergetics, what exactly is he referring to?

Nick Lane: He's referring specifically there to Huber and Wächterhäuser's 1997 paper showing that starting with carbon monoxide and methyl sulfides you can produce things like acetyl phosphate or at least methyl thioacetate, which is a thioester, and that was catalyzed by metal ions (Ni) and metal sulfides (FeS and NiS) alone. And so starting with the kind of carbon compounds that he wants the vent to be giving, which is to say, carbon monoxide and methyl sulfides -- these activated methyl groups that he's talking about -- it's possible to drive everything from that. So that's what he means by that remark.

I would like to derive those methyl groups, methyl sulfide, etc., by the direct reduction of the interface in the vent. But from there on, I think it's the same chemistry that he's talking about so the distinction there is quite small.

The distinction with what Mike is saying -- they are also deriving methane from the vents but they're then using things like nitric oxide in the oceans to oxidize it to give a methyl group. So this is a distinct hypothesis which is assuming a much more oxidized early ocean than I would assume. But like I would do, it's assuming that it's being driven in the vent itself rather than coming from serpentinization.

Mike is looking to drive the formation of pyrophosphate within the walls themselves, whereas I would look to drive the formation of acetyl phosphate by the conditions that Bill is talking about but driven by the proton gradients.

They're all slightly different views of really rather a very small problem but a very difficult problem because it's the source of all the organic carbon. It's the source of all the energy, all the carbon, and these are directly testable questions. And we're all trying to test it in different ways.

Suzan Mazur: So the difference between your work and what Mike Russell and Elbert Branscomb are doing -- there's not a huge divide.

Nick Lane: Not at all, Mike is the pioneer of this field. We differ on details, but the overall conception comes from Mike.

What Elbert Branscomb was saying in requiring a larger energetic push and what Mike has always said is that it requires not just a proton gradient but a redox gradient as well. And that the redox gradient is providing an extra oomph of power which Bill and I do not have without the redox gradient. That's essentially what their argument is. He's saying we also need to have strong oxidizing agents -- nitric oxide or things like that in the ocean. And then you have both the redox gradients, you have hydrogen inside and nitric oxide outside – there's a lot of energy in that couple. On the one hand. So you also have a proton gradient, pH 11 inside and pH 5 or 6 on the outside. There's a lot of energy in that as well.

My question about that is, the trouble is most of the organic chemistry that people talk about works very well in anoxic conditions. So getting hydrogen to react with CO2 and producing amino acids, etc. -- it's all favored under anoxic conditions. Now I think they're playing a little bit fast and loose with what anoxic means. They're saying there's no oxygen there but there is nitric oxide, which in terms of its reduction potential is very similar to oxygen. So they're claiming really microoxic conditions in effect. Those themselves should inhibit the forward reaction to make organics. Now they would say it will work if you have a compartmentalized system and they may be right, but I think they're calling on something which is quite difficult to really imagine for me. So I would want to do without it.

The reason being that to drive carbon reduction for me requires a continuous flux of protons across membranes or thin inorganic barriers -- it is no use if the barrier retains an electrical potential over many hours. It needs to dissipate in seconds, but be replenished in seconds too. If it's the case that protons flux that rapidly into the compartments, then nitric oxide would too. And if that's the case, then the compartments would be microoxic in their redox potential, so further carbon reduction to form amino acids, etc., would no longer be favored.

[Bill Martin has responded:

> "Thanks for the kind words, and that's fine! There are clearly many different opinions out there on how much energy was needed to get life going. I think that harnessing the outside energy sources was an invention of the biological world, and the ubiquity of A- and F-type ATPases among genomes would certainly attest to that. Being a biologist, I think that microbes are the authorities when it comes to defining the thermodynamic limits of life. Microbes like acetogens and methanogens that lack cytochromes live at the thermodynamic limits of life. How they do that is fascinating. They abundantly use inorganic (metal) catalysts under strictly anaerobic conditions to catalyze reactions that are very similar to those going on at some hydrothermal vents today. They harbour the kind of enzymes and reactions that I think are required."]

Suzan Mazur: Nigel Goldenfeld recently told me the following regarding life:

"Once we understand why the thing can exist in the first place, then we can understand how it is instantiated in any particular system."

Michael, are you saying you now understand why "the thing can exist" and "how it is instantiated" -- that it is somehow "baked in"? How do you envision life kicking in, how does the I-am-in-charge moment happen?

Michael Russell: For my own private happiness, I think we have an idea of how life starts. I think it would start that way on any other planet because these wet, rocky worlds always have the same kind of ingredients, the same kind of disequilibria, the same kind of free energy gradients. So, I'm quite happy about that. The only difference in the biochemistry would be at the cellular level, at the bacterial level, in what's called chirality, that is, the handedness of the molecules, that they could be possibly a mirror image of the structures of our DNA and proteins. But overall life will always do the same job -- it hydrogenates carbon dioxide from the air or dissolved in water with hydrogen released either photochemically or geochemically from water to produce a small but ever-renewed amount of organic molecules. Only at the beginning would it require geochemical/hydrothermal methane.

Wittgenstein famously said, "Don't ask for meaning, ask for use." Basically life does a job for a world that's out of equilibrium. So to me it's kind of pretty straightforward, and I say we've got a good outline. I admit we don't have all the dots, but we feel we can trace it right through now from the geochemical environment, which is very far from equilibrium right up to a bacterial cell. We recognize the stages, but of course, there are big gaps in our knowledge and understanding.

I come from a geological point of view and think in terms of standing on a canyon side and looking across the canyon at established life as seen by the microbiologist. There are researchers on the other side and I'm trying to "meet" them. We recognize huge similarities in terms of bioenergetics and the requirements for inorganic elements. We've got some stepping stones. They're rather far apart, but the stepping stones go right across the canyon. It's exactly opposite. As Schiller reportedly said, "and in the chasm lies the truth." Life employs so many things available at the submarine alkaline hydrothermal springs such as metals and phosphorus, etc.

Suzan Mazur: You're talking about sparking life. You're going from something being "baked in," something being imposed on a system and then that system comes alive, the system takes charge. How is that leap made? How does that happen?

Michael Russell: The kind of imagination I would bring to bear on this is -- that when you let out the bath water, within about six seconds, billions upon billions of water molecules cooperate with each other going down the vortex, down the plug hole. That's called self-organization. The universe itself gets very quickly into ways of self-organizing when there are large gravity or electrical gradients or whatever. The universe organizes itself. To me life is not such a peculiarity, it's complex but in a sense it is similar to the other mechanisms that get rid of -- or lower the energy gradients (increase the entropy) of the universe.

Suzan Mazur: So Nigel Goldenfeld's call for a general theory of life is what's needed because we don't quite understand what life is.

Michael Russell: I would take the philosopher view that it's what things do that really matters. You don't know me very well and I don't know you very well. But it's what we say that matters. We may never know each other that well. It's the same with all phenomena in the universe. We approach phenomena by use. We can see what they do. We can see the effect they have. I look at it from that point of view and then I try and understand the nitty gritty of life and its origin once I have that sense of it.

Suzan Mazur: **It's more relational than I-am-in-charge.**

Michael Russell: **Yes, it's relational. Absolutely.**

Chapter 10

The Geneticists

Trumpeting 'Life in Lab'

JACK SZOSTAK
(photo, courtesy Jack Szostak)

June 3, 2014

"You heard it here first," announced astrophysicist Mario Livio as Harvard biologist and Nobel laureate Jack Szostak told a World Science Festival "Search for Life" gathering on Saturday in New York that he expected to make "life in the lab" in three to five years -- and more likely within three years.

Livio dubbed Szostak: "The Leader of Origin of Life Studies in the Universe."

Szostak, however, did not go into detail as to why he thought he was so close to a synthetic cell that self-reproduces and cautioned that, at the moment, there are still lots of gaps in our understanding about a continuous pathway from chemistry to life.

Said Szostak, "For the parts that we do understand, sometimes we can see that there are two or three or four different ways this could have happened, maybe in slightly different environments."

Szostak described early-Earth conditions as still poorly understood:

> "In particular, we know that biology uses iron-sulfur compounds a lot to catalyze reactions. So why is that? Is it because that's something that was important in the beginning of life, or is it because that's just really good chemistry that life grabbed hold of later on and used for its own advantage?"

He raised the question about where the materials that make the building blocks of biology came from:

> "It probably can't all have come from the same environment. . . . It might be that some molecules, like what we make our membranes out of, might have come from deep hydrothermal vents. Other molecules might have come from atmospheric chemistry. Other molecules might have had to come from chemistry catalyzed by mineral particles. . . . Those are some of the things we just don't know that need more explanation."

Harvard Origins of Life Initiative Director, Dimitar Sasselov, also on the panel at the WSF "salon," told the audience that years ago origins of life was a neglected question, mostly because people thought it was too hard. But outside the meeting room he confirmed that they are indeed now "very close" to life in the lab. Sasselov thinks it will take five years, not three.

[**Note**: More recently, Sasselov told me he is "deferring to Jack" on the time estimate.]

This is fascinating considering that protocell development is roughly only a decade old.

Sasselov and Szostak, aside from their day jobs at Harvard, are both coordinators of the Simons Foundation origins of life collaboration, with Gerald Joyce (Scripps Research Institute), John Sutherland (MRC Laboratory of Molecular Biology - Cambridge), Matthew Powner (University College London) and others among the investigators, and with a team of postdoc researchers. Simons Foundation was one of the principal sponsors of this year's WSF.

In one lighthearted panel exchange Mario Livio described Jack Szostak as "a very competent person," adding, "If Jack manages to do life in his lab--"

"Then you know it's really easy," Szostak jested.

Part 4

Circus Toy Models

"In talking about the model I have fallen into a trap. I begin to talk about it as if it were historic truth. It is of course nothing of the kind. . . . It is only a toy model. . . ." -- **Freeman Dyson**, *Origins of Life*

Chapter 11

How to Make a Protocell

Jim Simons, Impresario Extraordinaire

JAMES HARRIS SIMONS
(photo, courtesy J.H. Simons)

"I remember you," said James Simons, origin of life philanthropist, math genius, 60s peacenik and so much more, searching my face as he greeted me in his office at the foundation he chairs in Manhattan. Indeed, we had met months earlier at a springtime lecture on von Neumann machines the Simons Foundation hosted downstairs in its state-of-the-art theater, and Simons invited me to stop by for an interview for my book -- with the caveat that he knew nothing about origin of life.

Our interview was scheduled following his summer at sea and return from the Far East. Last August, Simons addressed the International Congress of Mathematicians in South Korea, sailing there on his yacht, Archimedes.

We met again in September, and I was delighted when he decided to share his perspective on origin of life with me.

It was the day after Jewish New Year -- Casual Friday. Simons was dressed in an open-neck shirt and khakis. He wore a copper-silver-onyx bracelet on his right wrist.

Simons is a little shy of six feet, although a back problem disguises his true size. A light beard does not disguise lips some have compared to Bogart's.

The office was empty except for his closest assistants. I was struck by the composure of the setting. White shades transform the view from the foundation's windows overlooking the Flatiron district, and there are none of the usual distractions: No photos of Simons shaking hands with political leaders, cigar-smoking generals or revolutionaries. No sports trophies. Not a single ancient artifact in sight.

At first glance, the centerpiece of the foundation's executive suite of rooms seems to be a series of modern canvases painted in fuchsias, yellows, blues, greens, along the wall leading to the chairman's office. But it is his Chern-Simons equations which have significantly influenced theoretical physics, particularly string theory, that Simons has framed inside his office.

Chern-Simons is a quantum field theory in three dimensions first set out in a paper he co-authored with Shiing-shen Chern in the mid-70s: "Characteristic Forms and Geometric Invariants."

Forbes magazine lists Jim Simons as the 93rd richest human with an estimated wealth of $12.5B. The enormously successful hedge fund he chairs -- Renaissance Technologies -- enables his philanthropy.

Last summer the fund's success came under Senate scrutiny. Simons is a big contributor to progressive politics. This particularly irks Republicans, as well as some Democrats who argue that Uncle Sam could make good use of more of the fund's money. (For more war, perhaps?)

Understandably, Simons doesn't look for a lot of publicity. He seems to have nothing to prove, is at peace with himself. He's handled the tragedies in his life -- loss of two sons -- by reaching out to others through philanthropy.

The first and biggest Simons Foundation investment so far -- $100M -- has been for autism research, spurred by his interest in helping people like his daughter, who has a mild form of autism.

Simons finds it curious that historically some of the greatest scientists are now thought to have been in the autism spectrum. He doesn't think any stigma should be attached to it. "It takes a certain kind of talent to wall oneself off for a while," he told me, "to mentally isolate."

Jim Simons mesmerizes in conversation. He takes you into his world -- yes his eyes are deep, profound -- where he calmly plays with ideas until they have flesh, then delivers with an irresistible bounce. Like other great minds, he's great fun.

A slice of Simons' humor can be found on YouTube in the what-it's-like-to-be-a-billionaire interview he gave not long ago to a black entertainment television show. The conversation is staged with Simons seated beside the desk of his host in what looks like a middle school classroom.

"I'm Jim Simons, this is *Celebrity High*," he says, sporting a gold Rolex and two gold rings (one is his wedding band) and revealing that being able to afford dinner tops the list of his most favorite things about being rich.

Simons also enjoys asking surprising questions during interviews like, "How old are you?" -- and trading stories. So we traded stories. I recounted one of mine about the US bicentennial fashion show I modeled in on stage at the Royal Tehran Hilton in 1976 with former CIA Director Richard Helms attending the gala and dancing up a storm in his official capacity as US Ambassador to the Shah's Iran. (www.scoop.co.nz/stories/HL0409/S00244.htm)

Simons laughed when I told him the show followed Spiro Agnew's appearance. (Agnew had resigned under a cloud three years earlier as Richard Nixon's vice president.)

Jim Simons is a native of Brookline, Massachusetts, the only child of a shoe factory owner, and he still has a trace of a Boston accent even though he's been a resident of New York for decades.

He began his career at age 14 working in the basement stock room of a garden supply store. Simons says he was soon demoted to floor sweeper, a job he liked because he could think about math, and about girls.

Along the way he married two substantial women. His wife Marilyn Hawrys Simons holds a PhD in economics from SUNY - Stony Brook and has been the architect of the Simons Foundation. His former wife, Barbara Simons, is an award-winning computer scientist with a PhD from the University of California -- Berkeley.

Jim Simons received his undergraduate degree in math "early" from MIT and then moved on, on scholarship, to study math at Berkeley, where he was awarded a PhD at age 23 -- for his dissertation: "On the Transitivity of Holonomy Systems." By the late 70s, Simons was amassing a fortune trading foreign currencies, having left behind the Stony Brook math department he built and chaired not only in better shape but one of the best.

Prior to Stony Brook, Simons taught at MIT (he's now on the board of trustees), Harvard and Princeton. In the 1960s while at Princeton, Simons was also a cryptanalyst at IDA (Institute for Defense Analyses). It was at the time of the Vietnam War and he did advanced code breaking for the National Security Agency. Everything was "hush hush," Simons says.

He told the South Korean Math Congress last summer (www.youtube.com/watch?v=RP1ltutTN_4) the story about his decision not to stay in the shadows on Vietnam. Simons said the head of IDA, General Maxwell Taylor, fired him after he advised Taylor he'd given an interview to *Newsweek* in opposition to the war and written a letter to the *New York Times* as well saying war was stupid.

"The only available course consistent with a rational defense policy is to withdraw with the greatest possible dispatch." Simons wrote. The *Times* published Simons' letter as a followup to its magazine cover story in which Taylor said the US would soon win the Vietnam War.

Simons' Vietnam-era wisdom still seems to resonate. He told me he thinks everyone should be able to have their own beliefs and live happily together. "It works in New York," he said.

He also thinks it's important for people to do something beautiful and original in life and not give up.

Jim and Marilyn Simons' generosity has indeed been beautiful. Aside from the Simons Foundation support for autism research, it has donated $60M to Berkeley for a theory of computing institute; $150M for medical research to Stony Brook -- where the Simons both have roots -- plus $60M for creation of the Simons Center for Geometry and Physics at Stony Brook.

We now have a National Museum of Mathematics (MoMath) thanks to the foundation, which has also given $50M to Math for America. Roughly $80M so far has been pledged to the Simons Collaboration on Origins of Life. The foundation is a sponsor of the World Science Festival as well.

This just scratches the surface of the philanthropy. Jim and Marilyn Simons have signed the Giving Pledge, meaning they will give away most of their wealth.

My interview with Jim Simons on origin of life follows.

September 26, 2014
Simons Foundation, New York

James Simons: I had been thinking about life in the universe. How long it might have been with us -- it didn't start on Earth/ didn't it start on Earth -- but more generally, how it might be distributed through the universe.

I was in Belgium to give a talk and I met Martin Rees, who's a great astronomer and a friend. He was there. I started quizzing him about carbon, because carbon was not created in the Big Bang --

if there really was such a thing. Only the lighter elements were. In fact, it [carbon] was the lightest element that was too heavy to have been created. The heavier elements got created when the first supernovae, these exploding stars, fused lighter elements and made heavy ones. That's how heavy elements got here. That was about the year 300 million, as I seem to reckon from the Big Bang.

I said to Martin, "So I suppose carbon's been around from almost the beginning -- 300 million years out." Because I figured life -- we know of no life on Earth that's not carbon-based. It's not that it doesn't depend on oxygen or this or that, but it all seems to be carbon-based.

But he said, "No, no." He said it would have taken another two billion years before the concentrations of carbon would be high enough so that you would see anything. You had to wait until there were galaxies before carbon might be in sufficient quantity to do something.

And then he asked me if I was interested in origins of life. And I said, "Yes, that's an interesting subject."

Suzan Mazur: This was when?

James Simons: I don't know. Three years ago maybe. I think about three years ago.

He [Martin Rees] is very interested in this, which I didn't realize, and he waxed eloquently, and he pointed out that it's not very well funded.

I said, that's very interesting that it's not very well funded. So he sort of put that in my head. Then just about two years ago or a little bit more, in June, we had a two-day conference in a place called Buttermilk Falls, upstate. A little bit upstate on the Hudson. We had 20 great scientists there of all stripes from all fields to talk about collaborative projects. Goal-oriented collaborative projects that the foundation might do.

For a long time, it [the foundation] did one big one [project] -- autism. But I was thinking of smaller ones, still sizable, where there was a goal and would involve collaboration.

Jack Szostak was one of the presenters there. Origins of life had been in my head as something we might want to focus on. Jack gave a great presentation. One thing led to another. That was the first project that we started that came out of Buttermilk Falls. Since then I think we've done five. But that was the first one. That's how it started.

Suzan Mazur: Their Harvard Initiative [the Szostak, Sasselov *et al.* Harvard Origins project] was already underway for a few years. But it needed more--

James Simons: It needed a lot more. Yes. It needed a lot more. In fact, I think they were maybe even losing funding. I don't recall. But this [the Simons Collaboration on Origins of Life] is on a substantially bigger scale. And it's a great question.

Suzan Mazur: Is it the puzzle that intrigues you?

James Simons: Well I think everyone.

Suzan Mazur: You're being very generous in your support. There must be something very compelling.

James Simons: I think it's a compelling question. It's just a question that everyone would like to know the answer to. Will we get the answer while I'm still alive? I don't know. But I think we'll make progress towards it.

I think of it as a plausible path to RNA. That's the way I see it. Once you had RNA, well then it should be easy, relatively speaking. Evolution could do its thing. Cells would get together and form multi-celled creatures. But RNA was a pretty important spot to get to, and there are probably lots of plausible paths. But so far we don't even have one plausible path.

Suzan Mazur: Jack Szostak made an announcement at the World Science Festival that he hoped to have "life in the lab" in three years.

James Simons: He could. But that's not going to answer the question of the origins of life. Maybe shed some light on it.

Suzan Mazur: But at that point, suppose he does have "life in the lab," do you think then Wall Street would be interested in more direct investment?

James Simons: I think from life in the lab, as Jack might create it, would be very, very far from an application. But I don't know what he would create, so I can't answer that question.

Suzan Mazur: What do you think the value of making a protocell is at this point? What do you think its possible applications could be? Have you thought about this at all, discussed it with some of the other scientists?

James Simons: I guess that's a part of this project. There are all different places along the way -- it could start from rocks. You could say okay, how did cell membranes form? Jack is an expert on membranes. How did a cell form? Earlier it would be, just how did we get RNA? DNA, but RNA seems to have come first. So there are different places.

Basic science done by outstanding scientists, you never know where it's going to lead. You never know what, if anything, it's going to apply to. But surprisingly, often it does apply to something. So if you ask me -- Jack's very good, I'm sure his work is being done with good taste. Will that apply to making drugs or who knows what? I don't have the faintest idea.

Suzan Mazur: But he [Szostak] is very clear in his statements about the collaboration. He's not doing this single-handed.

James Simons: Yes.

Suzan Mazur: He consults colleagues--

James Simons: Sure.

Suzan Mazur: -- who are part of the collaboration and other people. It's a big undertaking.

James Simons: Yes, but it's not the central focus. The collaboration has many foci. People looking at different stages of this. Dimitar [Sasselov] is looking at exoplanets.

Suzan Mazur: And he's making a UV light for the protocell project.

James Simons: Who is?

Suzan Mazur: Dimitar.

James Simons: Dimitar. I didn't know that.

Suzan Mazur: It takes a certain special light.

James Simons: But Dimitar is an astronomer not a cell biologist. He's mostly focused on things that astronomers would likely be focused on.

Suzan Mazur: Are you aware of other philanthropies supporting origins of life research? Aside from yours and Marilyn's Simons Foundation. Harry Lonsdale's [Origin of Life] Challenge.

James Simons: I know about him.

Suzan Mazur: Also Jeffrey Epstein has supported Martin Nowak's work at Harvard going back a few years. I think that's ongoing. According to Dimitar Sasselov, Nowak is still supported by the Epstein Foundation.

James Simons: Could be.

Suzan Mazur: Do you know of others?

James Simons: No. I don't.

Suzan Mazur: Do you think it's a good idea for other--

James Simons: I think it's a very interesting field. I think it would be fine if others got into it. Sure.

Suzan Mazur: What do you think the national security interest is in synthetic cell development?

James Simons: I don't have the faintest idea.

Suzan Mazur: DARPA [Defense Advanced Research Projects Agency], for instance, is supporting a lab in the Midwest.

James Simons: Yes. Well, DARPA is involved in chemical and biological warfare, for example. Certainly protecting us from that sort of thing.

Suzan Mazur: Protection.

James Simons: Could be. But I think you have to ask someone at DARPA what their idea is. I don't really know. They have a far reaching mandate. A lot of scope in their mandate.

Suzan Mazur: Funding in Europe seems to be somewhat drying up for synthetic cell development, particularly in Italy. As a matter of fact, one of the key [Italian] officials, I believe the prime minister, was on *Charlie Rose* last night complaining about how private companies were not supporting very important aspects of the culture, including technology and science.

James Simons: Private companies as opposed to foundations.

Suzan Mazur: Well, the private sector. Philanthropy also doesn't seem to be happening in Europe for this type of investigation.

James Simons: Philanthropy doesn't happen in Europe very much period. There's not a very, very big. . . Philanthropy in the United States, *per capita*, is far greater than it is in Europe. We have a very old tradition of philanthropy. There is philanthropy in Europe but it's not done on the scale that it is here.

The government, as it does here, supports a lot of the research in Europe. Private enterprise used to do more of it in the United States, in particular. Big companies, most notable would be Bell Laboratories, RCA had a big facility -- where there was a continuum. Most of the funding would be for research that was really close to being applied. But way down -- down there somewhere -- there was a good deal of just plain basic research.

When I got my PhD, the best place in the world to study what's now called condensed matter physics, and in those days it was called solid state physics, was at the Bell Telephone Labs. That was a better place than Harvard or MIT or anyplace else. They did the best research.

Suzan Mazur: [Rockefeller University physicist] Albert Libchaber worked there, on and off.

James Simons: A lot of people did. He might well have. I know a lot of people who worked at Bell Labs. It was a remarkable institution. But when the telephone company was broken up, when AT&T was broken up, none of the individual companies really had enough money to so generously support that kind of enterprise.

But other companies -- IBM probably doesn't support as much basic research as it used to. And neither does the federal government, for that matter, because they're cutting research support in general. The NIH is down about 10% or more in real terms. NSF is getting cut. And more than that, they're being pushed more towards applications. Translational research. "Let's see some results tomorrow."

Suzan Mazur: Cost-effective.

James Simons: Right. Cost-effective. You know, the average Senator is not so interested in something that might happen in 20 years as a result of someone's research. He wants to be able to tell his constituents, "We supported this and look."

There's less patience, and basic research requires a lot of patience. You don't know, really, what's going to come of it. Remarkably or maybe not so remarkably, as happens -- typically, it does move downstream to applications we didn't know were going to be.

Suzan Mazur: As an extraordinary mathematician, what is your perspective, what is your thinking about whether life is algorithmic or non-algorithmic?

James Simons: I don't even know what the question means. What do you mean by that question?

Suzan Mazur: Is it a question of physics? Is it something we can figure out with math? Or is there too much that's unpredictable for it to be algorithmic?

James Simons: Well, physics, yes. It's a physical phenomenon, so therefore, physics will certainly play a role. What is the physics?

You know, chemistry is a specialized aspect of physics. What was going on, let's say, if life originated on Earth, what was going on in the early chemistry of the planet might promote the kinds of molecules that would ultimately turn into what we call life. . . .

There surely is physics. And it may be that we'll discover that, although I don't think we'll do it in my time -- that certain general principles may have a tendency to create what we call life in a wide variety of environments. That's possible. I don't know what those principles are. I've seen some pushes in that direction.

Suzan Mazur: Are you talking about the exploration of active matter? There are inorganic chemists working on this sort of thing -- to make matter come alive. Not involving genetics.

James Simons: Way before genetics. Way before genetics. Although there might have been some precursor.

Look, the two things that characterize life are reproduction and metabolic activity -- metabolism. We create energy, trees do it, ants do it. We're taking in energy from somewhere and converting it into something else. At the same time, what we think of as life is something that does reproduce. I think those are the two characteristics.

Which came first? It's up for grabs. The consensus is that reproduction perhaps came first before metabolism. On the other hand, there could be some very primitive notions of metabolism that predated reproduction. I don't have any opinion whatsoever on that.

Suzan Mazur: But you don't think it's a 2D world. You think there's more to it.

James Simons: I don't think that it's a 2D world.

Suzan Mazur: You think it's clearly a 3D world?

James Simons: I would never have thought that until I saw that article that you just handed me. ["Do we live in a 2D hologram?" -- *Science Daily* re Fermi Lab's investigation]

I continue to think that it's a 3D world, and string theorists would tell us no, it's a 10D or 11D world. I'm pretty comfortable with 3D right now. Could it be 2D and we're all living in a hologram? I don't know. Maybe we'll all wake up and find out we've been having a dream. But probably not. . . .

Suzan Mazur: Your commitment and Marilyn's to science and to the public is extraordinary. We are all enriched by what you've established at the Simons Foundation over two decades. Thank you.

James Simons: Thank you.

Jack Szostak

JACK SZOSTAK
(photo, courtesy Jack Szostak)

June 23, 2014

There is a certain hush when the name Jack Szostak is mentioned in science circles. Yes he's a Nobel Laureate, but the pause seems to be more for a man who is trusted. He's also a handsome man, with fine Polish features (his paternal grandfather was from Cracow). The elegance he and his wife brought to the Nobel events, captured in photos, says more.

Szostak's autobiography mentions that his father Bill, an aeronautical engineer, built a chemistry lab for him in the basement of their home in Montreal during his high school years, encouraging and sometimes participating in the scientific exploration. Vi, his mother, supplied the "remarkably dangerous chemicals" from her job -- the mixing of which, he notes, "frequently led to explosions."

Szostak graduated from high school at age 15 and entered McGill University that same year. He earned a BS in cell biology from McGill and then a PhD in biochemistry from Cornell University.

I was first introduced to Jack Szostak's work in 2008 during a conversation with astrobiologist Bob Hazen, who liked what Szostak was doing with self-organization in synthetic biology. ISSOL (International Society for the Study of the Origin of Life) president Dave Deamer later steered me to Szostak's "great molecular animations."

Jack Szostak is a professor of genetics at Harvard Medical School; Alexander Rich Distinguished Investigator, Massachusetts General Hospital; Investigator, Howard Hughes Medical Institute; and a professor of chemistry and chemical biology at Harvard University. He also directs a lab at Harvard where he is researching origin of life and synthesizing a protocell -- from scratch.

Other big names in science, like Matthew Powner and Irene Chen, found a home early on in the Szostak lab as postdocs. Powner told me he was inspired by the "free-thinking" environment there.

Szostak, with Harvard astrophysicist Dimitar Sasselov, also coordinates a team of origins scientists funded by the Simons Foundation, among them Gerald Joyce, John Sutherland, Matt Powner.

Jack Szostak is a recipient of the following prizes: U.S. National Academy of Sciences Award in Molecular Biology; Hans Sigrist Prize; Genetics Society of America Medal; Lasker Award; Dr. H.P. Heineken Prize; and the 2009 Nobel Prize, with Elizabeth Blackburn and Carol Greider, "for the discovery of how chromosomes are protected by telomeres and the enzyme telomerase."

He is a member of the National Academy of Sciences, New York Academy of Sciences, and the American Academy of Arts and Sciences.

Szostak is author (with Dave Deamer) of the book, *The Origins of Life*, and of a large number of scientific papers. He is also author of a much-discussed commentary, "Attempts to Define Life Do Not Help to Understand the Origin of Life" (*Journal of Biomolecular Structure & Dynamics*), in which he says,

> "An inordinate amount of effort has been spent over the decades in futile attempts to define 'life' -- often and indeed usually biased by the research focus of the person doing the defining."

I first met Jack Szostak at the Simons Foundation this spring, and then again at the World Science Festival where he announced he'd have "life in the lab" in three to five years. Our conversation follows.

Suzan Mazur: What properties of life are you looking to synthesize in your protocell, and what are the outstanding issues in making the synthetic cell?

Jack Szostak: What we're trying to build in the lab is a protocell, a simple cell that can grow and divide, and, most importantly, start to evolve in the Darwinian sense. That requires several things.

With a cell model you need to have a cell boundary, a cell membrane that can grow and divide. We already have that, we've published on that. The other thing that is really critical is to have some kind of genetic material that can replicate so that useful information can be inherited, in other words, can propagate from cell to daughter cell.

What we're thinking about mostly these days is that that genetic material might be RNA or some closely-related material. One of the reasons we like RNA is that it acts even now as a genetic material. We also know that it can act as a catalyst. There are ribozymes, which are basically enzymes made of RNA.

Suzan Mazur: What are you missing at this point?

Jack Szostak: What we're missing is the ability to replicate RNA molecules without enzymes.

Suzan Mazur: What is the distinction between self-reproduce and replicate? Is there a distinction?

Jack Szostak: Some people make a distinction between perfect replication and replication with errors. Any kind of replication always has errors. To me the distinction is pointless.

Suzan Mazur: Is there another outstanding issue in making the protocell aside from getting it to replicate?

Jack Szostak: That is the big issue, getting it to replicate. That problem has a number of sub-problems.

Suzan Mazur: Norm Packard, whose company ProtoLife has worked on synthetic cells, recently told me this:

> "Right now, the way we see RNA getting produced is with this enzyme that catalyzes the production of RNA. But how do you get that enzyme? Well that enzyme is a thick, giant protein. So somehow you have to bridge this gap of getting RNA production to happen without this enzyme or create a story of how the enzyme gets created along the way. But so far this hasn't happened."

You're exploring non-enzymatic RNA replication.

Jack Szostak: That's right.

Suzan Mazur: So this continues to be the sticking point.

Jack Szostak: People have been trying to solve the problem of non-enzymatic replication for the last 50 to 60 years. The late Leslie Orgel and his colleagues and students, including Jerry Joyce, spent a lot of time on that problem. They made a lot of progress, but ultimately they got frustrated and concluded that it was impossible.

Everybody started to look at alternatives, either alternatives to RNA or different modes of catalysis. For example, RNA catalyzed RNA replication, which we spent a lot of time on. Now we've come back to the idea of reassessing those problems. We know a little bit more.

Suzan Mazur: But you've said you think you'll have "life in the lab" -- as Mario Livio put it -- in three to five years. So what kind of life do you think you'll have?

Jack Szostak: I might have been feeling optimistic that day. I think we've solved most of the issues on the way to RNA replication. The big one that's remaining is how to bring chemical energy into the system.

The problem is RNA falls apart. The activated nucleotides we use to do the non-enzymatic replication -- they react with water, so they fall apart. There needs to be a way to bring energy back into the system to essentially keep the battery charged. To keep all the nucleotides activated and to keep things running. If that problem can be solved, then I think we will be able to do non-enzymatic RNA replication.

And because we've solved a lot of the other issues, that will allow us to generate a replicating protocell.

Suzan Mazur: How collaborative an effort is this? Aside from your own team there at Harvard, are you looking to external labs like John Sutherland's--

Jack Szostak: Sure.

Suzan Mazur: For the mix that you're going to put into your lipid vesicles, is that right?

Jack Szostak: Yes, we definitely talk a lot to John and his lab in Cambridge and to Matt Powner, who was a student of John's and a postdoc with me. We talk frequently with Jerry Joyce and Donna Blackmond at Scripps. Basically everyone who's active in the field and thinking creatively about these problems, we all talk to each other and hope we can together come up with a solution.

Suzan Mazur: Are you referring to the Simons Collaboration on the Origins of Life?

Jack Szostak: Those people I mentioned are part of the Simons collaboration.

Suzan Mazur: What Matt Powner told me a few weeks ago was that he and John Sutherland were able to come up with "robust and high yielding synthesis for two of the four nucleotides, uridine and cytidine" and "we'll get to the purines as well and maybe DNA."

Dimitar Sasselov, your co-coordinator at the Simons collaboration told me that synthesizing the protocell will require special ultraviolet light, which he's making.

It sounds like there's a bit more work to do.

Jack Szostak: The problem of the origin of life, if you stand back and look at the whole thing, is a whole pathway all the way from planet formation to the various chemical steps like making nucleotides, to assembling a protocell -- which is what we're working on -- and then on to the evolution of the genetic code, etc.

What I meant when I said three to five years was, given the building blocks, if we have the building blocks, we'll be able to generate an evolving protocell. There may still be gaps in our knowledge of how prebiotically to make nucleotides or maybe gaps with what happens afterwards. We're just working on that little part of how you get the molecules to work together and act like a cell.

Suzan Mazur: We may never know precisely the origin of life, but how much insight into the actual origin of life do you think a protocell like yours might give us, say on a scale of 1 - 10?

Jack Szostak: We've been working on one part of the problem. Other people are working on other parts of the pathway. I don't worry at this point whether we'll know exactly how it happened on the early Earth. What we're trying to do is to work out a plausible pathway where all of the steps seem chemically and physically reasonable, and maybe we'll end up with multiple pathways which are all possibilities.

Suzan Mazur: Aren't there a half dozen or so labs working on some type of synthetic cell right now, all with different approaches?

Jack Szostak: Most of the other people who are working on artificial [synthetic] cells are taking existing biological cells and trying to take them apart to reconstitute them from components, or they are trying to simplify them.

Suzan Mazur: Minimal cells.

Jack Szostak: Right.

Suzan Mazur: What applications might result from a successful protocell synthesis?

Jack Szostak: At the moment I can't see any practical applications. To me it's just intellectually interesting.

Suzan Mazur: What are the potential dangers of unleashing "Generation II," as Sasselov describes it? Could the cell have the robustness to interact with the environment outside the lab?

Jack Szostak: The things that we're trying to make, even if we're successful, will be totally dependent on an environment that will only exist in our lab. There is no place on the modern Earth like the early Earth.

Suzan Mazur: Do you think we need more public awareness about origin of life and what the plan is regarding the synthetic cell? People are curious as to why you're doing this. Yes, it's a fantastic intellectual exercise, but there must be some sort of vague goal.

Do you think the public should be better informed about it, maybe by way of a series of roundtables on the *Charlie Rose Show*, underwritten by one of the sponsors of the World Science Festival, or Pfizer, perhaps, like the Paul Nurse 13-part series on *Charlie Rose* a few years ago?

Jack Szostak: I think there's already been huge public interest. Almost everyone would like to know how we got here. With all the excitement of NASA and the exploration of our solar system and the detection of planets around other stars -- almost anybody who has any interest in the world we live in knows about that. I think that's already stimulated a lot of interest about the origin of life.

We do try to do things like public lectures. Being on TV shows takes a certain talent. Not everybody has the television charisma of Paul Nurse.

Suzan Mazur: When we met at the Simons Foundation in April you told me that you "don't believe" in autocatalytic sets. Why is that? Haven't the Europeans integrated autocatalytic subsystems into their systems science?

Jack Szostak: Autocatalytic sets is one of those concepts where the people who came up with the original idea, like Stuart Kauffman, rather than admit being wrong kept changing their story until it was basically the same concept everybody was already working on.

The original idea was that there would be large numbers of compounds where one would help another to replicate, and that one would help some other one to replicate, and that somehow out of this huge population of interacting molecules autocatalytic replication would emerge.

In my opinion that was never chemically realistic. Now you see people talking about non-enzymatic RNA replication and calling that autocatalytic sets. If that's what you want to call it, that's fine. But it seems like the concept has lost all meaning.

Suzan Mazur: I see. Thank you.

Dimitar Sasselov

DIMITAR SASSELOV
(*photo, Jon Chase, couirtesy D. Sasselov*)

"We told Harvard from the start [in 2006] that there is no such thing as an origins field. There is astrobiology, and there are individuals working on origins-of-life experiments, and what we want to do is not really fundable any other way. We told the provost committee, "We need you to share our vision and support our research because the federal agencies -- NASA, NSF, NIH -- won't." We added that we also couldn't predict whether it would work out". **-- Dimitar Sasselov**

July 28, 2014

Astrophysicist Dimitar Sasselov is director of the Harvard Origins of Life Initiative -- a group of scientists that includes Jack Szostak, George Whitesides, Andrew Knoll, Martin Nowak *et al.* -- and co-director (with Jack Szostak) of the Simons Collaboration on the Origins of Life, which says a lot about the confidence the scientific community has in Sasselov's leadership in this cutting-edge area. He's also a warm, engaging man with a talent for creative organizing in a field that astrophysicist Piet Hut once described to me as "herding-cat science."

One of Sasselov's most important contributions in science could turn out to be his expertise with ultraviolet radiation. Sasselov is now designing the UV light in which the world's first synthesized protocell, which he calls "Generation II," is expected to emerge in roughly three years and begin a new tree of life.

Dimitar Sasselov was born into a prominent science family in Bulgaria during the years of Soviet influence. He was raised in Nessebar, a town on the Black Sea coast founded by Greek colonists in antiquity, which Sasselov says is still rich in archaeological artifacts.

In fact, Sasselov's father, also named Dimitar, is a dirt archaeologist as well as an architect -- retired now and writing books -- with an interest in late antiquity and early antiquity (Thrace) through the late Roman Empire (Byzantium).

Sasselov's mother's profession is horticulture. She was responsible for the design of some of Bulgaria's parks.

Dimitar Sasselov said he was very much encouraged professionally by his parents and might naturally have chosen a career in marine biology, since his family lived near the sea. "But life is a string of accidents along the way," he told me, "so I ended up in physics and astronomy."

Sasselov found living with all the political red tape in communist Bulgaria "very frustrating." "Luckily, by the time I was in my early 20s," Sasselov said, "the whole political system was falling apart. It collapsed eventually in 1989."

I first met Dimitar Sasselov and his wife, an artist, last winter at the Manhattan apartment of Gerard Senehi, the mentalist who likes to host intellectual salons. Sasselov was guest speaker.

After about 10 to 15 minutes into Sasselov's talk about the origins of life and the search for "super-Earths," I, of course, wanted to know how Sasselov defined "life."

He responded by saying there was no agreed-upon definition. He has since told me the following:

> "There is clearly the realm of atoms and small molecules to consider, which is chemistry, and there is also something that emerges under conditions we are still trying to figure out. This is also chemical but has a level of organization making it qualitatively different. Therein lies the challenge.
>
> Where do we draw the line between those two extremes and generalize it, not just for microbial life on Earth, which I agree is the main representative of the life phenomenon, but also for other environments and with other combinations of molecules and chemicals? . . . I'm investigating the phenomenon from the point of view of the cosmic large scale.
>
> If we study the universe at large scale, we see the hierarchies that developed over the last 14 billion years, starting with very diffuse gas made entirely of protons, electrons and a small number of alpha particles, which will become helium. So it's hydrogen and helium, a certain amount of photons and gravitons, gravity waves, and other perturbations which are in the space-time matrix, the space-time that all this exists in and is

225

expanding. From that early time to today, different objects form -- galaxies, and clusters of galaxies. Within galaxies stars form, and around the stars, planets. . . . The point is: Where is the place of life in this whole hierarchy?"

Dimitar Sasselov is a professor of astronomy at Harvard University. He did his undergraduate studies at Sofia University in Bulgaria, where he also received his PhD in physics, and then received a second PhD in astronomy from the University of Toronto. He was a postdoctoral fellow at the Harvard-Smithsonian Center.

Sasselov is the author of the recent book *The Life of Super Earths* and co-editor (with Mine Takeuti) of the books *Stellar Pulsation* and *Pulsating Stars*.

He is a frequent and lively lecturer and knows how to handle the media, greeting me at the Senehi salon by telling me that he loved my blog.

Excerpts of my interview with Dimitar Sasselov follow.

Suzan Mazur: You've identified three milestones of *Homo sapiens* in your book *The Life of Super Earths*: the Copernican revolution, globalization, and synthetic biology. You note that the first two of these are already done deals. Here's what you say about the third, synthetic biology:

> "For the first time in about 4 billion years a new species is not going to emerge from the set of processes that led to the diversity of life on this planet. Instead, one species is going to synthesize another -- a life-form that is unique, but not in the way that a new dog breed or a genetically modified corn plant is made unique by some cosmetic differences with its progenitor. It will be new in terms of its unique biochemistry, a new life-form that has no place on Earth's tree of life, a new life-form at the root of a new tree of life."

My question is why have scientists become intelligent designers of life? What is your goal in creating "Generation II," as you call it?

Dimitar Sasselov: One, it's a very practical thing. If we want to understand the phenomenon, if we want to understand and study its history and the possible ways it appears out there in the galaxy, the easiest way is to build proxies or actual toy models -- as we usually call them in science -- in the lab. If we want to understand generally what would be the sensitivity or the reaction of a simple living system to changing planetary conditions, we can't use existing life forms, because they are too sophisticated. Hopefully we'll be able to synthesize a simple synthetic cell and see the feedback from experimenting with it.

It's fascinating that we are now going to be able to synthesize life in the lab because we think we've understood what the molecular building blocks are for over half a century. But we still don't have a sense of what makes them work as a system. We can't figure that out any other way than to build a synthetic cell.

Suzan Mazur: What is the goal beyond figuring out the synthetic cell?

Dimitar Sasselov: If the project is successful and the synthetic cell functions and produces populations, then there is so much to do with it, with those populations and this new tree of life. It will open biology in a completely different way. It will transform the face of science altogether.

You have to see what you get. But I'm sure it will happen soon. For me it's feasible. I know it will work.

We're doing the bottom-up approach. Chemist John Sutherland is very much involved in this, working from his lab in Cambridge. We were collaborating with John even before the Simons Foundation funding. The ultimate goal is to go all the way from his prebiotic chemistry, building the individual pieces of RNA and the other molecules needed, and then encapsulate them in Jack Szostak's lipid vesicles.

Suzan Mazur: In the various scientific collaborations now in progress on the origins of life, including the Harry Lonsdale-funded teams, would you say investigators are being cautious enough about the dangers of conspiring -- that is, making one person's research fit snugly into another's for a desired result? For example, if one had glycerine and the other had purines, then that would open the door to the monomers hypothesis, which could segue into your idea, etc., etc. Would you comment?

Dimitar Sasselov: I do think there are multiple pathways, but we can't make a statement in one direction or the other. One of our goals is to answer that question. Our approach is to look for possible pathways, not for *the* pathway. But even if we just find one, we'll be more than happy.

Suzan Mazur: So if we get a few more creation myths along the way, that's part of the process?

Dimitar Sasselov: Yes.

Suzan Mazur: Jack Szostak announced at the 2014 World Science Festival that we'll have life in the lab in three to five years -- closer to three. You said five years.

Dimitar Sasselov: I'm deferring to Jack here.

Suzan Mazur: You began to describe the synthetic cell you're making there at Harvard as a modest protocell.

Dimitar Sasselov: Modest, yes. It should be built from scratch in order to understand each piece and function, as opposed to using it as a black box, which works but you're not quite sure why it works.

Suzan Mazur: From your perspective, what's missing from the soufflé at this point?

Dimitar Sasselov: There are two steps missing out of the eight steps toward an RNA protocell that Jack and others are still working on. One of those steps not yet figured out is how to make RNA strands grow longer and grow generally in order for the individual pieces -- called RNA nucleotides -- to bond into a strand. The RNA nucleotides have to be chemically activated. This is easy to do in the lab but tricky for it to happen naturally. That's one of the issues.

The other one is enabling RNA to form -- without any initial templating -- in lipid vesicles that are floating in a clay-water solution containing the chemicals necessary to form nucleotides. Templating is using a strand chemically as a photocopier to make more copies.

These two things should come together in the next three years. That's why Jack said at WSF that in three years we hope to have all the components together and have vesicles with activated nucleotides in them.

These components will follow the chemical steps we know from doing those steps individually in the lab. But now the components will be a system, and for the first time we'll have something similar to an evolution arms race -- Darwinian evolution -- between the individual vesicles with their particular selected RNA strands. Hopefully we'll see it go very quickly through several generations, see the selection process working.

Suzan Mazur: Thank you. I understand you're exploring the interaction of radiation and matter, that you're trying to recreate the light of prebiotic Earth in your lab. What was different about the light then? How far along are you with the experiment?

Dimitar Sasselov: All this interesting chemistry and biochemistry that we've been discussing, including and especially including the synthesis of the building blocks, is now increasingly well understood. John Sutherland and Jack Szostak and George Whitesides and others have managed to make a lot of progress in that direction. One pervasive aspect of the pathways -- it's not a single pathway but several pathways -- is that they always require ultraviolet light for different non-trivial steps in the chemical reactions which are involved.

That means that we have to model and actually recreate this UV radiation environment in the lab in order to do the whole sequence properly. As it happens, this is my expertise. I'm building a small lab where we'll be doing all of that.

So technically the integration of all the different steps discussed above will eventually have to be done in the lab I'm putting together, where I'll be simulating conditions under UV radiation, either corresponding to the early Earth or to some of those exoplanets we're discovering and will be discovering and studying.

Suzan Mazur: The light of prebiotic Earth was stronger.

Dimitar Sasselov: Yes. There was not the oxygen there is today in the Earth's early atmosphere. Earth's oxygenation was subjected to harsher, stronger UV light.

Suzan Mazur: John Sutherland did say at last year's annual Lonsdale meeting that he doesn't have the proper light (and couldn't afford to get one). He said he's using 254 nanometers.

Dimitar Sasselov: Yes, chemists work with mercury lamps or other lamps which produce a spectrum of UV light completely diametrically opposed to what a star would produce in terms of UV light. It still works partly because what UV light does is simply deliver energy.

A lot of what the chemistry requires is the right kind of UV radiation as a source of energy. There are some subtleties there, particularly in terms of left-handed versus right-handed, and that still unsolved problem of polymerization (*i.e.*, the activation and growing into a longer strand), which I think will be solved with the proper UV source.

Matthew Powner

MATTHEW POWNER
(*photo, courtesy Matt Powner*)

June 17, 2014

Protocell pioneer Matthew Powner has the fresh-face good looks of an athlete you'd expect to show up at World Cup 2014, and a voice reminiscent of the "British Invasion" (he's from the north of England). But the relaxed focus he projects signals something profoundly more urgent. Matt Powner's got the chemistry of origin of life on his mind. Even though he says he actually does love to play a bit of football now and then.

Nobel Laureate Jack Szostak, who recently told a World Science Festival audience that we'll have "life in the lab" in three years, once described Powner as "already a star" by the time he arrived as a postdoc at his Harvard lab five years ago because Powner, as a PhD candidate at the University of Manchester, and John Sutherland, then his advisor, were able to find a way to synthesize two of the four nucleotides of RNA -- uridine and cytidine. Szostak considers this one of the pivotal events in origin of life research. And indeed, a crucial part of what remains to be done to meet Szostak's announced three-year deadline depends on the creative chemistry of John Sutherland and Matthew Powner.

In 2012, Sutherland and Powner teamed up to win Harry Lonsdale's Origin of Life Challenge and $50K plus $150K in research funding. Powner has now established his own lab at University College London where he continues his origins investigations. Sutherland works from MRC Laboratory of Molecular Biology - Cambridge.

Aside from renewed funding in 2013 and 2014 from Lonsdale -- Powner and Sutherland are now also being supported as investigators at the Simons Collaboration on Origins of Life, headed by Jack Szostak and Harvard astrophysicist Dimitar Sasselov.

Matt Powner's BA, MS, as well as his PhD in chemistry are from the University of Manchester.

He is the recipient of numerous honors and awards, including, as mentioned, the Lonsdale Origin of Life prize shared by John Sutherland; Roscoe Medal (Science Engineering and Technology for Britain); Stanley Miller Early Career Research Award (International Society for the Study of the Origins of Life); Royal Society of Chemistry Prize (University of Manchester); Swan Prize (Univeristy of Manchester); Merck Sharp and Dohme Award (University of Manchester), among others.

I met Matthew Power recently at the Simons Foundation in New York. Our interview follows.

Suzan Mazur: Were you at all divided in your career choice in science? Was there something else that you were considering?

Matt Powner: When I left undergraduate study, I first interned at AstraZeneca. I thought my career in science meant making drugs,

233

working for a pharmaceutical company, because I had this passion for organic chemistry and building molecules. But following the AstraZeneca internship, I decided I wanted a PhD before actually working in Pharma. While looking for a PhD, I found John Sutherland and made a U turn into origins of life research.

As an organic chemist, investigating origins seemed valuable, intellectual, something I really wanted to put my time to for three years or so. From that point on there's been no other course I've really thought about pursuing.

Suzan Mazur: How far back does your passion for science go?

Matt Powner: I don't think I was sciencey at school. I was just quite good at it.

Suzan Mazur: Did you have parallel interests, music, acting -- the arts?

Matt Powner: Sports. I did a vast amount of karate as a kid. And other sports -- football.

Suzan Mazur: Acting?

Matt Powner: No. Science just really made sense to me. I'm not sure where it comes from, good teachers and perhaps the way I interacted with my parents. The sciences just appealed to me -- all of them -- physics, chemistry, biology. I could get along with those subjects more so than the arts.

Suzan Mazur: You have an interesting accent.

Matt Powner: I come from the north of England.

Suzan Mazur: Do you come from a science family?

Matt Powner: Not really. Dad has been a farmer and a plumber and electrician. Mom's background is banking, math-related but not science.

Suzan Mazur: Were you always a scholar?

Matt Powner: I always did very badly at school until nearly the end of high school. In fact, I had a chemistry teacher who told me chemistry just wasn't the subject for me.

Suzan Mazur: I first learned of your work on origin of life two years ago when you and John Sutherland won Harry Lonsdale's research prize. Would you highlight what your original proposal was and where you are now with that Lonsdale-funded research?

Matt Powner: In 2011, John and I published a paper suggesting that in a search for a global system that could be geologically plausible -- that could give rise to multiple systems essential to life -- we should couple chemical research and the investigation of chemical systems.

We used the example of nucleotides and lipids. Our thinking was that if we could find multiple systems to cross-reference chemically so that one type of chemistry could give rise to both those classes of molecules, we could use that to infer what geochemistry was at life's origin.

Our proposal was to use chemistry to define geochemistry rather than the usual origins research approach of beginning with a model of geochemistry to define the parameters of the chemistry.

Suzan Mazur: Where would you say you are now in your research on this?

Matt Powner: Funding to our labs by Harry was renewed last year at the same level based on peer review by Ram Krishnamurthy, Jack Szostak, Irene Chen and Nicholas Hud. John

Sutherland's lab is at Cambridge and my lab is at University College London.

We've submitted a one-year report to Harry Lonsdale and his reviewers and we gave an online seminar. They were happy we'd made progress. We are about to give our second-year report.

Funding was also renewed at the same level to the other researchers originally awarded grants by Harry: Niles Lehman, Peter Unrau and Paul Higgs; and David Deamer's team -- Wenonah Vercoutere and Veronica DeGuzman.

Suzan Mazur: Can you talk a little about the results?

Matt Powner: I do think we've made a lot of progress over the last two years. Some of the work, specifically from John pushing this, is looking at copper cyanide chemistry and reduction of simple molecules, reduction into cyanide to give sugars. So, linking chemical systems to photochemistry, which really kind of resembles the way biology operates now.

Suzan Mazur: Is the Lonsdale funding to support your research and John Sutherland's research in general or is this a separate project that's being funded?

Matt Powner: It's in general. It's specific to our labs.

Suzan Mazur: Is that the same situation with the Simons Foundation funding you and John Sutherland are receiving? Simons is supporting your research in general?

Matt Powner: Exactly. The Simons Foundation supports the lab and research program of each investigator as well as specific research proposals of each of its postdocs.

The Harry Lonsdale money, which is great, supports half a studentship -- half the cost of one PhD student for my lab.

236

Suzan Mazur: I know the Simons Foundation postdocs are being funded at $50K plus another $30K for health insurance, travel and supplies. Are the investigators receiving a larger sum?

Matt Powner: The investigators get different sums depending on how senior they are.

Suzan Mazur: The origins research is being done in labs outside the Simons Foundation in all cases?

Matt Powner: Yes, I'm pretty confident that it's all outside the foundation. I think the biggest strength of the Simons Foundation Origins Collaboration is that the group is really broad. This inspires people to think outside the box about everything from astrophysics -- with people like Dimitar Sasselov involved -- from planets, stars, galaxies, right down to single molecules and how they work, to the atomic scale and everything in between.

Suzan Mazur: I interrupted your research description.

Matt Powner: Part of our research relates to the making of very simple molecules. Taking hydrogen cyanide and building more complex 2 and 3 carbon species. Another aspect is looking at the next level of complexity. How we can build a nucleotide, something that's significantly more complex -- although it's a long way from a biological species -- how to build bigger macromolecules from that. We're doing quite well, I think, on all fronts.

Suzan Mazur: Do you define your work as making a protocell or are you exploring how it could be made? What is the goal?

Matt Powner: We're interested in aspects of development all the way to a protocell. We're a smaller, younger lab than John's lab. At the moment our focus is molecular compositions, an earlier stage. We're looking at core components of biology and why they may have formed the specific way they have and how that can be controlled by chemistry.

Suzan Mazur: The end goal is to make a protocell?

Matt Powner: Partly.

My end goal isn't just to make a protocell. You could make a protocell any way you want to. You can manipulate all the parts. None of it could look biological. You could make membranes of non-biological polymers. You could have non-biological information transfer. You could have something resembling metabolism that looks nothing like biology and that will tell you a lot about assembling something that can act and look like a cell.

The focus in my lab now is understanding why biology took the route it did. Biology has assembled a specific set of components. We want to shed light on why THOSE specific components through the rules of chemistry.

Suzan Mazur: A protocell can be built at this point, but not one that can self-reproduce.

Matt Powner: Some labs would say they can build a protocell. But, yes, getting biological tight self-sustained replication is [down the road].

Suzan Mazur: Is John Sutherland's lab more aggressively pursuing the making of a protocell?

Matt Powner: I guess you should ask John.

Suzan Mazur: Well his work has been described as "research to discover systems chemistry, the syntheses of the informational, catalytic, compartment-forming molecules thought necessary for the emergence of life."

Matt Powner: Yes.

Suzan Mazur: Sounds like protocell development.

Matt Powner: To me it sounds like he's trying to understand the chemistry that can assemble the parts you need for a protocell.

Suzan Mazur: And then you can make one?

Matt Powner: Information of how you assemble those into a protocell is not clear. Whether John's lab is currently looking at the assembly of those systems, which they could well be, you need to really ask John.

Suzan Mazur: Prior to your winning the Lonsdale Origin of Life Challenge, I understand you were a postdoc in Jack Szostak's lab in the US. Were you working on protocell development there? What can you say about your experience in the Szostak lab?

Matt Powner: I worked more on the genetics side in Jack's lab. I was there for just over two years, 2009 - 2011. I started at University College London in 2012.

Jack's lab at that time had largely two aspects. One side researched protocells, lipids, assembly, growth and division of vesicles. I worked on the other side of the lab, the genetics side, where the focus was genetic molecules, genetic polymers. We played more with RNA, the natural component. We also moved a little bit away from RNA and explored non-natural genetic polymers.

Suzan Mazur: How would you rate your experience there?

Matt Powner: The Szostak lab was a really free-thinking place, great to work in. Lots of bright people to sit and talk with. Lots of freedom to explore my own ideas as well.

Suzan Mazur: How much of your work there was experimental and how much computer simulation?

Matt Powner: It was 100% experimental.

Suzan Mazur: What about your work now?

Matt Powner: Same thing, 100% experimental. We only use computers to draw pictures of what we did in experiments. Powerpoint and Word is about all we use computers for.

Suzan Mazur: Jack Szostak has commented that you were already "a star," "incredibly focused" when you came to his lab as a postdoc because of your work on synthesizing nucleotides with John Sutherland. Can you tell me about the nucleotides work and also your research on purines as building blocks of DNA and RNA?

Matt Powner: My initial project when I joined John's lab was to look at TNA (threose nucleic acid).

It was one of Albert Eschenmoser's ideas for non-natural, non-endogenous nucleotides to propagate information that may have been important in the early stages of life. During an in-depth discussion with John, John and I decided to remodel my project to attack the problem of ribonucleotides.

RNA is a cornerstone of modern biology, a cornerstone of genetics, inheritance, and evolution. We weren't willing to accept that RNA was inaccessible at the dawn of evolution. It seemed so important to us, yet synthesis was still a huge stumbling block for RNA, something scientists essentially agreed was critical at some early stage in evolution.

So I started thinking about what was known about nucleotides and what the best routes out there were and how we could change them, come at them from a slightly different angle.

John and I used as a starting point some of the vast amounts of work done by Leslie Orgel and colleagues, and without Orgel's previous work what we did would have been infinitely more difficult.

240

What we were able to find, without getting too technical, is that by slightly changing the conditions, by building our molecules in a slightly different order -- the order in which we made the bond in the final nucleotide product -- we could come by really robust and high yielding synthesis in two of the four nucleotides, uridine and cytidine.

That was important in reinvigorating RNA as not the only player in town but as a really important part of biology. Those syntheses suggest that it may have been possible that it was a really important molecule in the first biology. The control of chemistry could allow you to synthesize at least those two components of RNA and we'll get to the purines as well and maybe DNA. So we found that the specifics of fairly simple chemical reactions could in a controlled manner give these two nucleotides.

So there's always been lots of very intriguing chemistry, catalysis, etc. that relates to large pieces of RNA but the real stumbling block was how to get the original monomers that you can synthesize, long pieces of RNA that can then interact.

I'm not necessarily saying you have to have monomer-by-monomer synthesis of RNA, but to make RNA you're going to have to make a pool of nucleotides.

Suzan Mazur: Can you discuss what's missing from the mix at this point? What are the outstanding issues in creating a self-reproducing protocell?

Matt Powner: We'll need 10-20 years to make the kind of protocell that actually progresses toward biology, one that could be described as a missing link between prebiotic chemistry and what we now know as life. That is in some ways distinct from just any protocell.

You could envisage lots of variable components of a protocell that relate in no way to biology. Subsystems of a protocell, such as membranes, for example. But what's missing for me are the interactions between the components, like how you get

a hereditary molecule to interact with the membrane-forming molecule so those start down the path toward coupled evolution -- beyond assembly -- understanding how the system could self-assemble from chemistry.

Much more needs to be understood -- like coming up with a reasonable purine synthesis. A lot of this relates to building bigger and bigger systems, bringing together multiple aspects of molecular entities and getting them to interact in a way that's productive.

Suzan Mazur: Is autocatalytic sets a somewhat marginalized approach at this point in protocell development?

Matt Powner: I don't think the idea of autocatalytic sets has been marginalized, but I don't know if it's necessary. My understanding of the concept is that autocatalytic sets can in essence themselves evolve purely through change in chemical composition. I'm not sure that's been demonstrated in a relevant system that doesn't rapidly degenerate, and I'm not sure that's the essential step to building what we know as a modern cell. However, if we actually found that autocatalytic sets work, this could be a lynchpin to understanding origin of life.

People should research what they think is important. Scientists will think different things are important and that's good. You don't want everybody to focus on the same question or part of the question. What we will almost certainly come to realize in the fullness of time is that everyone was at least in part wrong but multiple theories were in some small part right. Without a crystal ball the only way to move things forward is through empirical evidence.

Suzan Mazur: Why build a protocell in the first place? I think you told me when we met at the Simons Foundation that it was largely sort of getting to the bottom of things, the puzzle was intriguing.

Matt Powner: Yes. We're fundamentally interested in the scientific question: Why is nature the way it is.

Suzan Mazur: And beyond that, no one knows until it's made what the applications could be?

Matt Powner: From that technology can flow. But none of it means anything until we know the results, what it now requires is getting self-assembly at this level to occur.

But you never know what blue skies research is going to turn up. And the increasing emphasis on systems chemistry could be hugely advantageous for reinvigorating the whole field of chemistry as well as the economy.

Suzan Mazur: There are some grumblings that it's one of NASA's missions to control origins of life research. Would you comment?

Matt Powner: As someone who's never been involved with NASA, I don't feel restricted by what NASA does or by what NASA says. I don't feel that NASA is controlling origins of life research and I don't think they should. I'm not sure to what end they would want to either.

Suzan Mazur: What about the European space program, is there a NASA counterpart there on origin of life research?

Matt Powner: No, there's no European Space Agency exo-biology space program like NASA has.

Suzan Mazur: Why do you think that is?

Matt Powner: Level of funding.

Suzan Mazur: Origin of life is funded in a different way in Europe?

Matt Powner: Yes -- just my guess.

Suzan Mazur: Are you in favor of more public funding and more public awareness regarding origin of life research or do you think the subject is too esoteric?

Matt Powner: Origin of life is a problem no one knows the answer to. But we can formulate a concept of a sterile planet. We can formulate a concept of a biosphere. Anyone can do both of those things. What links those two and how we could move from one to the other is hugely interesting.

I think the public wants to know what's happening with origin of life developments and should know. It is something that increasingly inspires people in science, and it is rare in terms of projects where chemistry can have a large input.

Where we came from is a fundamental question, compelling for everyone irrespective of background. I think there should be more public awareness of developments, and increased public funding could actually follow. But this is a loaded question for someone in the field.

Chapter 12

How to Make a Minimal Cell

Vincent Noireaux

VINCENT NOIREAUX
(photo, courtesy V. Noireaux)

It was positively springtime in Minneapolis. Raining lightly, glistening, a bit windy. Following my non-direct flight from New York, the air was like a rush of Peter's Blend (a serious Bleecker Street roast).

I was excited by Vincent Noireaux's invitation to visit his lab at the University of Minnesota, his minimal cell lab. Noireaux offered to meet me at the airport. But since I'd never been to Minneapolis, I decided to make my own way into town, to explore.

Vincent Noireaux is 40ish, quick, with attractive blue eyes and an unmistakably-French smile. His PhD is in physics from France's Institut Curie.

As a postdoc, Noireaux collaborated with Albert Libchaber on a synthetic cell system, working in Libchaber's condensed matter physics lab at Rockefeller University. Libchaber is formerly director of research at CNRS (French National Center for Scientific Research).

We met in Noireaux's office. In the grass outside his building are a pair of condensed matter sculptures Noireaux introduced me to, recent additions to the university landscape called *Spannungsfeld*, meaning "tension field."

The pieces are the creation of German quantum physicist-turned-artist Julian Voss-Andreae, now an Oregonian (or "Orygunian," as they say). The 10-foot tall male and female torsos each weigh 1.5 tons and are made of sliced steel and open space. Viewed from various perspectives, the figures appear to change form, from solid to hologram.

Voss-Andreae says the sculptures are "a metaphor for the counterintuitive world of quantum physics."

(*Spannungsfeld* photo follows.)

(Spannungsfeld photo courtesy of artist Julian Voss-Andreae, photo by Patrick Siegrist)

Inside Vincent Noireaux's office I notice a baby carriage. Noireaux informs me that his wife is a scientist as well. She's now finishing her PhD in physics at the University of Minnesota.

Over a Starbuck's coffee, Noireaux tells me he grew up in the French countryside. His father is a government official and his mother a banker.

On the day of my visit, Noireaux was in the middle of a move to the university's 3M building, to a spacious new lab, which I toured -- featuring state-of-the-art equipment and windows looking out into the Minnesota sky. It's a minimal cell research facility largely funded by DARPA (Defense Advanced Research Projects Agency).

But first we took a spin through Noireaux's old cramped and windowless *E. coli* lab, the door posted BIOHAZARD.

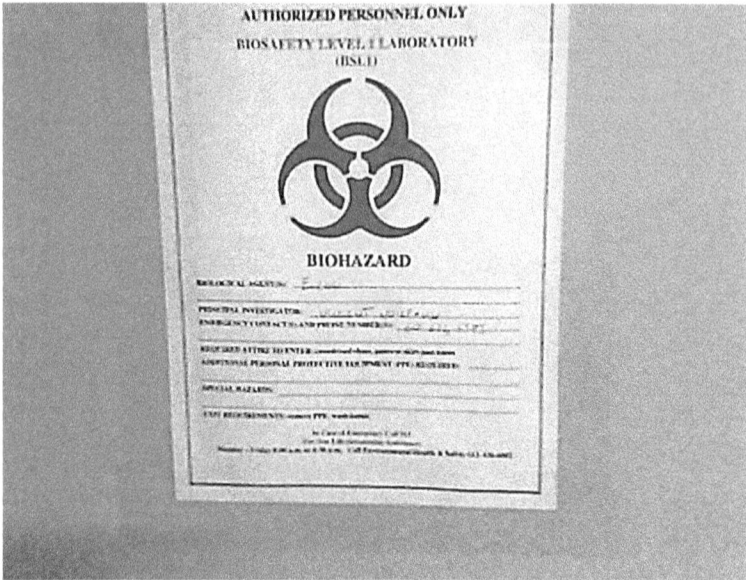

(photo by Suzan Mazur)

Test tubes filled with fascinating liquid were being readied for transport. Noireaux noticed that a floor freezer was a bit ajar, and he quickly put a make-shift weight on top of the lid to close it. We next passed an innocent-looking white refrigerator that Noireaux opened to reveal shelves of tubes of much less innocent-looking stuff. We then proceeded out of the lab.

I didn't want to spend more time than necessary around the tubes, frankly, considering the biohazard sign. But I did wonder what DARPA's interest was in Noireaux's minimal cell project.

Noireaux says even he doesn't know what he might discover with his minimal cell research.

Noireaux works with a team of graduate students and one postdoc, Filippo Caschera, who he introduced to me in the room adjacent to the old lab. Caschera was busily organizing the move. He'd been a researcher in Venice, Italy at Norm Packard's company

ProtoLife before coming to work with Noireaux and at Pier Luigi Luisi's minimal cell lab in Rome.

During our interview Noireaux described his current work on the minimal cell and also mapped out who some of the other synthetic cell players are.

May 12, 2014
University of Minnesota

Suzan Mazur: Can we walk through the three types of synthetic cells? And would you tell me at what stage of development you are with your system?

Vincent Noireaux: I am now finishing a review with my postdoc associates in the lab. What we see are the following three approaches to making a synthetic cell.

First is the protocell, which is about origin of life. It involves creating a self-reproducing unicellular entity with the basic molecules of life, including molecules for the membrane, ions, some genetic information like RNA.

The second approach is the minimal cell. The minimal cell is made of natural molecules and components. It's more sophisticated. Machineries such as transcription and translation are used to execute the DNA programs, to try to create a self-reproducing unicellular system. It's known now that approximately 200 - 400 genes are necessary to make a minimal cell that would be able to reproduce itself.

The third approach is the artificial cell. I define artificial cell as a synthetic cell system that incorporates non-natural molecules and components. I can give the example of block copolymers, which look like phospholipids and self-assemble into membranes. Eventually it is thought artificial cells can be made from them.

In my laboratory here at the University of Minnesota, we're working on minimal cells. We have developed a cell-free transcription-translation system to execute large DNA programs *in vitro*.

Suzan Mazur: Cell-free.

Vincent Noireaux: It's *in vitro*. There's just no cell there. We extract the molecular machineries (for transcription and translation) to express DNA programs. We developed recently what we call the cell-free transcription-translation toolbox, which allows us to express relatively complicated DNA programs.

The largest one we've executed so far is a natural DNA program. We just did a demonstration to challenge the system with the genome of a virus, a virus from bacteria -- bacteriophage. This system is approximately 60 genes. It's still not the 200 to 400 genes needed for a minimal cell, but we demonstrated that the system we developed can take or execute very large DNA programs.

This electron microscope image is a picture of the system. You can see the phages that are made after a few hours entirely synthesized from the genome of these viruses. So we really have a system now. It's a liquid solution, where you put some DNA in it and the DNA is expressed. We are able now to make very well defined living entities, even though it's a virus and needs a host. It's a complex, self-assembled information-based system.

Suzan Mazur: Do you consider a virus living?

Vincent Noireaux: It's a very good question. It does require a host. It reproduces through a host. So whether it's living or not is a little bit debatable.

We have developed this system, which we call a cell-free transcription-translation platform, that allows us to execute DNA programs *in vitro*. We are returning now to the minimal cell. We want to encapsulate the system into liposomes and execute DNA programs that encode for essential functions of living cells.

What we do is take free cells and extract the complex machinery to express DNA. We remove everything, all of the genetic information of the cells, and we have all of these molecules we then use to express DNA that we synthesize in the laboratory. We can encapsulate that into liposomes.

250

Suzan Mazur: In trying to get it to self-reproduce, how close do you think you are? And what is the problem at this point in getting it to self-reproduce?

Vincent Noireaux: That's a very good question. The problem is. The cell is made of three parts [draws diagram]: information, metabolism, and there is self-organization. Each of these parts is made of molecular machineries, each of these parts is essential to make a cell.

The problem we have is to understand how these parts talk together. The real problem is the integration of these three parts in a compartment. Information is the DNA program, its genetic composition and its regulation. Metabolism is energy or the nutrients, at least at the beginning. And self-organization occurs with expression of proteins, which are able to make very specific assemblies and structures in specific locations. It is a real problem integrating these three parts in a container and coordinating their actions.

I have been working a lot on information and a system to express DNA. In the past few years, this has developed nearly to a system.

We've also worked to develop a system that can express protein for a very long time based on energy. Our platform is now more robust energetically.

Finally there is self-organization. We're trying to understand how when some proteins are expressed together, it is possible to make very specific structures.

In a sort of test before the minimal cell, we are looking at phages. In a certain amount of time, minutes or hours, you get something that is incredibly well defined. There are many things related to cells that we understand with this system, which recapitulate all of the fundamental steps of genetic information and its expression.

So you have information, you express a bunch of molecules, and they create something *perfectly* well defined. The phage in this case is a crystal, it's a crystal with DNA inside it.

251

The ultimate goal of what we are doing is the minimal cell, but first we have to understand the relationship between information and self-organization. Where we are right now is that we have the most versatile and powerful cell-free transcription-translation system reported so far for synthetic biology applications.

Suzan Mazur: So time-wise where are you with development of a minimal cell that can self-reproduce?

Vincent Noireaux: It may be early to give an estimation of how many years. We have a system which we think is relatively close to a minimal cell.

Suzan Mazur: How does recent creation of an organism with an expanded genetic code intended for use in the drug industry affect your current research, if at all?

Vincent Noireaux: For now, this work is relatively far from what we do. The current concern for us is to develop genetic regulation that will coordinate the expression of the genes. How to make a synthetic genome that works.

Suzan Mazur: Your interest is also in the minimal cell for use in drug delivery, I understand. In light of current thinking about top-down evolution, that evolution happens top down -- tissues, organs, cells -- and systemically, that it is not gene-centered -- how do you think that plays out regarding human exposure to these future drugs.

Vincent Noireaux: If we get a minimal cell, we can engineer it for specific application such as drug delivery or use it as a factory to synthesize new drugs. First, what is the state of the art right now in building a minimal synthetic genome? It is still difficult to make genetic circuits of a few tens of genes with predictable behavior. So we are relatively far, in that sense, from the minimal cell of 200-400 genes.

Does this work have ethical problems. Is it dangerous?

Suzan Mazur: Yes, that's my question. Is it dangerous if it's used to produce drugs people are taking?

Vincent Noireaux: I think we are relatively far from such achievements. The synthesis of a real minimal cell that can self-reproduce presents more potential ethical problems than the applications that would come after.

Suzan Mazur: Could drugs made via the minimal cell using natural DNA prove more problematic once in the bloodstream than, say, a vaccine made using synthetic nucleotides?

Vincent Noireaux: Here I do not have the elements to answer this question.

Suzan Mazur: But if the minimal cell is used to produce drugs that people are taking--

Vincent Noireaux: Yes. You can think about that. Synthetic DNA really depends on what you do with it and where you express it.

My work is more about natural DNA in a container for the minimal cell. The first minimal cells able to reproduce themselves will still be relatively fragile mechanically. They're not going to have the real robustness of a cell as we know it. They're not going to be dangerous, in the sense that they will certainly not be competitive.

Suzan Mazur: What I'm referring to is the epigenetic factor -- evolution that's top-down and systemic -- organs, cells, tissues -- not gene-centered. What could happen with the introduction into the human bloodstream of these drugs?

Vincent Noireaux: Your point is good definitely. If we have a minimal cell, how quickly is it evolvable? How quickly can it diverge and become something very robust? That is completely unclear what can happen.

Suzan Mazur: It's down the road and would have to go through all kinds of trials.

Vincent Noireaux: Exactly. Trials and engineering after that. Absolutely.

Suzan Mazur: Who would you say the key protocell, minimal cell, artificial cell players are?

Vincent Noireaux: Protocell, I would say Jack Szostak at Harvard. Pier Luigi Luisi and all the family of Luisi, Pasquale Stano *et al.* at the University of Rome3 -- they do both protocell and minimal cell.

[**Note**: Pier Lusi Luisi advises the Luisi lab is winding down due to lack of funding.]

Vincent Noireaux: There's the group of Steen Rasmussen at the University of Southern Denmark. They do modeling and both protocell and minimal cell.

Suzan Mazur: I interviewed both Rasmussen and Luisi. I haven't interviewed the Japanese researchers, like Tetsuya Yomo at Osaka who's working on the minimal cell. Albert Libchaber mentioned a former student of his, Yusuke Maeda now at Kyoto University, also working on minimal cell and protocell. I saw Maeda's Princeton Origins presentation last year.

Vincent Noireaux: There is also Takuya Ueda at the University of Tokyo. Ueda does minimal cell systems. There's Sheref Mansy of the University of Trento in Italy, as well as Christophe Danelon in The Netherlands at Delft University of Technology.

Suzan Mazur: One of your associates here was formerly a researcher at Norman Packard's ProtoLife lab, wasn't he?

Vincent Noireaux: Absolutely. Filippo Caschera. He was a postdoc at ProtoLife and got his PhD at the University of Southern Denmark. He was working with Steen Rasmussen and others.

Also, there's Pierre-Alain Monnard at the University of Southern Denmark.

The third synthetic cell is the artificial cell. The artificial cell is when you have synthetic components. More soft-matter people. Daniel Hammer at University of Pennsylvania is one scientist working on the artificial cell.

Suzan Mazur: There is also the Simons Foundation and its research team on origins of life. Actually, I think Sheref Mansy is part of that collaboration.

Vincent Noireaux: Our work here on the minimal cell is related to the origin of life but on a slightly more sophisticated level where we really try to understand the minimum genetic system. It's also a question of physics, how physically to make a self-reproducing entity.

Suzan Mazur: Research on origin of life is being increasingly privately funded meaning the research can be done more quietly. However, less public scrutiny may be a problem when the research is origin of life/protocell development.

Vincent Noireaux: I don't think there is a danger here, major results would be published. It is nice and essential that foundations support basic science.

Albert Libchaber

ALBERT LIBCHABER
(photo, courtesy A. Libchaber)

It was a perfectly sunny day in April to be anywhere, and I was on New York's East River taking in the tranquility of Rockefeller University's greenery, its landscape inspired by Versailles and now seriously maintained through the support of the Mary Lasker Charitable Trust.

Albert Libchaber, the Parisian-born physicist, invited me to stop by for a conversation about origin of life. He met me in Founder's Hall, the ivy-covered building just beyond the main gate, and escorted me through a sunlit atrium to his office where he offered me a seat on the sofa, choosing a straight-backed chair for himself.

Libchaber is a soft-spoken, gentle man who radiates optimism. Libchaber attributes his perspective on life to his childhood as a Jew growing up under the Nazis.

I was particularly interested in Libchaber's collaboration with physicist Vincent Noireaux on the minimal cell. Libchaber was quick to set the record straight that it is his friend Noireaux's minimal cell project, but that, yes, he does advise.

A decade ago, the two working together in Libchaber's Rockefeller University condensed matter physics lab (Noireaux was Libchaber's postdoc for five years) found a way to break through the membrane of a lipid vesicle using *Staphylococcus aureus*. Once inside the vesicle, working with an *E. coli* extract, they were able to synthesize a green fluorescent protein.

Prior to his current role at RU as professor of physics and director of an experimental lab, Libchaber was a physics professor at Princeton University and at the University of Chicago. He also served as research director of CNRS (French National Center for Scientific Research) for almost a decade (1974 - 1983) and had a long association with Bell Telephone Lab before that.

Albert Libchaber's PhD is from École Normale Supérieure, University of Paris.

Some of his honors include: Prix des Trois Physiciens (Foundation of France); Wolf Foundation Prize in Physics; John D. and Catherine T. MacArthur Foundation Fellowship.

My interview with Albert Libchaber follows.

April 28, 2014

Suzan Mazur: Life is both chemistry and physics?

Albert Libchaber: Yes, and my interest lately focuses more about life originating in thermal vents deep in the ocean. In thermal vents you have large temperature gradients because water comes out of the vents at a high temperature.

The little volcanoes that make thermal vents are full of porous material, and those pores hold all the temperature gradients you need. What we showed is that with this temperature gradient many nonequilibrium processes exist.

For example, we discovered that polymerase chain reaction (PCR) is possible under thermal convection. There's nothing but thermal convection. In the center region of a convective cell double-stranded (DS) DNA melts and in the side single-stranded (SS) DNA is copied. We thus made the smallest PCR machine, centimeter in size. . . .

We did that when Dieter Braun was here as a postdoctoral fellow, about 10 years ago.

We also showed that if you have a very thin geometry, water cannot move because of friction to the wall but the suspension moves. So in a temperature gradient, you can accumulate the suspension.

There can be a huge accumulation. We saw that you can accumulate, you can amplify, you can even select length. So all those processes may have played a role at the origin.

This is one of the questions always asked. How do you reach a critical concentration? Well, temperature gradient can do that. It can also amplify, it can select size. It can select even some sequences.

Suzan Mazur: Do you think life started on Earth from scratch?

Albert Libchaber: That's a hard question. I'm just saying that if RNA is present, then there are processes. How does RNA come about? That's chemistry. That's where chemistry is fundamental.

Once you have a coded polymer, then we show that you can select size and some sequences. You can do many things. But how did this polymer come about? That's chemistry.

Suzan Mazur: What would you say the timeline is in making a minimal cell?

Albert Libchaber: It depends what you define as a minimal cell. Is it a cell that self-reproduces? Or is it just a cell that functions?

A cell which functions, we've made it. For example, what

Vincent Noireaux can do with his cellular extract is very powerful. But he has not been able yet to make a cell that self-reproduces, using *E. coli* extracts.

Self-reproduction is a complex process. I think my friend Noireaux will be able to do that within the next five years. That's my impression.

Suzan Mazur: And the synthetic cell from scratch? How long for that?

Albert Libchaber: What do you mean from scratch? What do you call scratch?

Suzan Mazur: Building it from whatever comes before RNA, etc. From bottom up. Freeman Dyson said give it a hundred years.

Albert Libchaber: Oh. I don't know. But, yes, this will take a long time.

Suzan Mazur: I asked Matthew Powner and he said 10, 20 years.

Albert Libchaber: More.

Suzan Mazur: How long have scientists been working on the protocell?

Albert Libchaber: About 10 years. But for me biology is a branch of physics which arrived through the possibility that the code is there, through the genetic code that will be followed by self-reproduction. When you have self-reproduction everything is exponential. You double and double and double. So in a very short time you multiply enormously. You are then out of physics.

Self-reproduction is a new concept. That is what may have happened. We try to understand this concept in a minimal cell.

Suzan Mazur: How many labs are there working on the minimal cell?

Albert Libchaber: Very few.

259

Suzan Mazur: Very few. A half dozen?

Albert Libchaber: Yes, because there's no support.

Suzan Mazur: Yours and Vincent's lab.

Albert Libchaber: And one of my former students, Yusuke Maeda works on this in Japan.

Suzan Mazur: Then actively working on the protocell from "scratch" you've got Jack Szostak, John Sutherland, Matt Powner. . .

Albert Libchaber: It's a problem of chemistry, I would say. Really chemistry. Chemists have to show how to synthesize. That's a tough question and one on which I'm totally incompetent. . . .

Suzan Mazur: Why isn't there any support for this kind of research?

Albert Libchaber: The only support in modern times is for medicine, medicine and medicines. Anything else gets limited support.

Suzan Mazur: Rockefeller is a private university.

Albert Libchaber: Yes. I have some research money. My research years ago was initially funded by a Japanese company. I tend to work with only one or two other researchers.

Suzan Mazur: Would you comment on the support the Simons Foundation is giving to origin of life research?

Albert Libchaber: They're just starting. James Simons is a mathematician, he's interested more in conceptual ideas. Biologists are not. The Simons Foundation is interested in conceptual ideas.

Suzan Mazur: Is origin of life something the public is not ready to hear about? What is the problem of getting public funding?

Albert Libchaber: The problem in modern times is that everything is aimed at application. That's the problem. Medical applications, technical applications. But not conceptual ideas. Conceptual ideas are ignored.

Suzan Mazur: But why make the minimal cell? Why make the artificial cell? I thought scientists were thinking that these cells will have applications.

Albert Libchaber: Everything has applications. But as scientists we don't work FOR applications.

Suzan Mazur: You won't know until you make the minimal cell what the applications can be.

Albert Libchaber: For me the idea is: Can I understand self-reproduction? This and the genetic code are the two essential aspects of biology.

Suzan Mazur: Are the origin of life camps still divided: the geneticists, metabolists, and compartmentalists?

Albert Libchaber: I'm a physicist. There are no camps. I'm not in any of those camps. I'm just exploring my own interests, which is trying to understand self-reproduction. . . . This is a key concept.

Suzan Mazur: In my conversation with Jack Szostak at a recent Simons Foundation event, I mentioned autocatalytic sets. He said he didn't believe in autocatalytic sets.

Albert Libchaber: He used the right word. Believe. He doesn't believe in them. Believe, that's a scientific word. It's ideology.

Suzan Mazur: But there have been experiments going on in this area. . . .

Albert Libchaber: In Germany, Albrecht Ott has experimented with autocatalytic systems. So when he says, I don't believe in autocatalytic sets, what does it mean? I don't understand this. You don't damage science by introducing a model. You just propose a model. It's an interesting model autocatalytic sets. If it's a reality, I don't know. . . .

The situation is that origin of life is still a discipline, not yet a science.

Suzan Mazur: But you're also thinking about the synthetic cell. You're still collaborating with Vincent Noireaux on the minimal cell?

Albert Libchaber: Yes, we are still collaborating. Vincent thinks he will have a minimal cell that can self-replicate within five years. But it's not my work, it's his work.

Suzan Mazur: Do you have any ideas about how this minimal cell will be used once it's developed?

Albert Libchaber: There are many possibilities, but I don't know.

Suzan Mazur: Do you think there's too much reliance in the field on computer simulation?

Albert Libchaber: It's a good tool, computer simulation. Like the autocatalytic is an interesting model.

Suzan Mazur: But you have to balance the computer simulation with actual experiments.

Albert Libchaber: Yes and experiments are costly. Currently there is not much support for origin of life experiments. Simulation is certainly a way of the future.

Suzan Mazur: Do you think life is algorithic or nonalgorithmic?

Albert Libchaber: Both. The genetic code automatically has an algorithmic effect. But there's a nonalgorithmic effect. That's the problem in biology. It's mixes discrete mathematics with continuous mathematics. The algorithm is part of discrete mathematics. . . Biology also involves geometrical models.

This is what von Neumann in the 50s understood very well. This is the complexity of biology. It's a mixture of the two mathematics. It's partly algorithmic. Not totally algorithmic. When the cell divides, there's a lot of geometry. And there's a lot of chemistry. And chemistry is not algorithmic. It's continuous.

Suzan Mazur: So with computer simulation you can only rely on it to a degree.

Albert Libchaber: Yes. And you certainly know that before Crick and Watson, von Neumann proposed a logic of self-reproduction in machines. In that he showed that you begin with a tape, then you have to copy the tape. When you have a Universal Constructor, it takes coded information from the tape and builds a machine. Now that part exists today in the 3D printer. The 3D printer is really an invention of von Neumann's in the 50s.

Suzan Mazur: It's incredible, isn't it.

Albert Libchaber: It's remarkable! It shows you how important conceptualization is, not now but 60 years on. A lot of times there is a rejection in science of conceptualization but conceptualization is important in science.

Suzan Mazur: Do you think about consciousness at all in your work. In your work on the synthetic cell, for example?

Albert Libchaber: It's too early. We don't have the base. It's too speculative. Consciousness is a very exciting problem on which I don't have much to say.

Suzan Mazur: Do you have your own definition for life?

Albert Libchaber: It's a mixture of algorithm and geometry. It's the only place where both come together. Algorithmic means that you need a code. You need coded polymers. Otherwise there's no life. And then as soon as you have a code, you have self-reproduction as a possibility. Because once you have a code you have a memory. And once you have a memory, you can reproduce.

And geometry because life is physical. It's a combination of both that makes life. So how does this combination come out, this is what we show in the lab. . . .

Suzan Mazur: Is the US moving in one direction and Europe moving in another direction with origin of life research?

Albert Libchaber: In general, the United States is a country that produces, that makes. That's the American genius. Europe is becoming like Greece, a philosophy place. . . . America doesn't care too much about philosophy. I like and respect conceptual ideas, but I am an experimentalist.

Suzan Mazur: It used to be that evolution happened once you had life. Now the thinking is that evolution happened before life. How far back does evolution go?

Albert Libchaber: I think that before it started there were many tries. One try succeeded.

Suzan Mazur: Do you have any concerns about the unleashing of the minimal cell, the synthetic cell? The discussion is we won't know what we've made until we've made it and won't know until then how it can interact with life as we've always known it.

Albert Libchaber: It can reproduce in an artificial environment. Not in a normal environment.

Suzan Mazur: But what about the protocells made from scratch? It's not going to happen for a long, long time?

Albert Libchaber: I think stem cells have more possibility of changing human life. Much more than the minimal cell in a positive and negative way.

Suzan Mazur: You have a marvelous calm overview.

Albert Libchaber: As a little Jewish boy, I was raised under the Nazis. I was without parents. It was not easy. I learned very early in life to put things in perspective. I've lived through wars and many difficulties since. Life goes on. We live now in extraordinary times where all the big countries, particularly China and the US, are in agreement. Europe also. This has never happened before. It's a very special time. . . .

Chapter 13

How to Make an Artificial Cell

Steen Rasmussen

STEEN RASMUSSEN
(photo, courtesy S. Rasmussen)

"[W]e as humans are naïve if we believe we are the end product of evolution. I'm sure we are not." -- **Steen Rasmussen**

I last spoke with physicist Steen Rasmussen this past summer while he was en route to his lab at the University of Southern Denmark. One of the things we touched on was autocatalytic sets (molecules said to catalyze one another's production). The subject was on my mind because I'd recently interviewed Nobel

Laureate Jack Szostak who said he didn't "believe" in autocatalytic sets, that autocatalytic sets were not chemically realistic (Chapter 11).

Rasmussen seemed annoyed by Szostak's pronouncement. An October 2014 paper of his in the journal *Europhysics Letters* spells out why. Rasmussen's virtual computer experiment relies on autocatalytic networks in looking for a precursor to life.

Rasmussen also seemed rattled by my inquiry about progress on the origin of life proposal for a collaboration with CERN as brand. He was one of the participants at COOL EDGE 2013 but has since backed off efforts to make the project happen. Rasmussen would not go on record as to why.

Following is my feature interview with Steen Rasmussen about origin of life.

December 7, 2012

Hal (*2001*)? Roy (*Blade Runner*)? *Terminator*? Dinosaurs of *Jurassic Park*? Danish physicist Steen Rasmussen thinks seriously about sci-fi scenarios most of us see as Hollywood entertainment. But then Rasmussen is the flag bearer of artificial life and was one of the founders of the AL movement in the 1980s, also organizing the first two international protocell conferences in 2003. His first laboratory? A farm in Denmark "chasing cows" as a kid, thinking about the stars, drawing and dreaming. . .

At Los Alamos National Laboratory where he spent two decades as a physicist, Rasmussen -- with finely-chiseled features and a sprinter's physique -- also flirted with acting, once portraying the late Robert Oppenheimer in a comedy about how the atomic bomb would be impossible to make there today because of excessive security and red tape.

He has dabbled a bit more in theater since his return to Denmark five years ago, but Steen Rasmussen's principal role is heading the University of Southern Denmark's Center for Fundamental Living Technology. His decision to return to Denmark was based on the availability of basic research funds not dramatic roles.

Rasmussen says he is fascinated by a technological future with "democratized material production," *e.*g, an at-home personal fabricator to spin out medicine, electronics, clothing, "anything." Maybe more importantly, he sees us inventing machines that can "love more deeply" than may be humanly possible.

He thought a decade ago we'd by now have developed protocells, but insists we are "very close" indeed to synthesizing one.

"For large parts of my life I may have been working 70-80 hours a week," says Rasmussen with a certain laugh signaling that science is also great fun.

If you're thinking Alec Guinness, *Man in the White Suit*, the ex-Cambridge scientist who succeeds in cooking up a fiber that never gets dirty -- well, yes, Rasmussen admits the recipe does require a pinch of obsession.

Aside from his above-mentioned role at SDU, Rasmussen is principal investigator for Denmark's Initiative for Society and Policy (based at SDU), whose mission is to give science-based knowledge more visibility in the public discourse "to counterbalance the ideologies."

He continues a quarter-century long affiliation with Santa Fe Institute, now serving as an external research professor following 16 years as a researcher in residence.

Rasmussen's career began at Los Alamos, first as a postdoc in the 1980s, then as the "Team Leader for Self-Organizing Systems," and head of the Protocell Assembly project. He was co-director of Europe's Programmable Artificial Cell Evolution project (PACE), principal investigator for "Cell-Like Entities" sponsored by the US Air Force (2004-2005), and co-principal investigator for "Water on Mars" at Los Alamos (2003-2005).

Rasmussen co-developed the Transportation Simulation System (TRANSIMS), which the US Department of Transportation later implemented. He also developed an integrated simulation framework for urban systems and web-based disaster mitigation tools once used to reconnect 20,000 people who had been evacuated, tools also employed following 9-11.

Steen Rasmussen is the author of hundreds of scientific papers and presentations, and co-editor (with M. Bedau, L. Chen, D. Deamer, D. Krakauer, N. Packard and P. Stadler) of the first book on protocells, *Protocells: Transitions from nonliving to living matter*. He was also featured in the *Nova* program "Life From Dust" and is a frequent public lecturer.

Among Rasmussen's many scientific honors are: P. Gorm-Petersens Mindelegat (awarded in the presence of Queen Magrethe II of Denmark); Los Alamos Cerro Grande Wildfire Award (2000); Los Alamos Achievement Award for Excellence: Protocell design, and for Simulation of Critical Infrastructures (2004); World Technology Network Award, Biotechnology Category: Protocell design.

Rasmussen's PhD is in physics from the Technical University of Denmark.

He says he is happy to be back home in Denmark and values the basic research opportunities Europe now provides, but that he really misses the spirit and openness of America's Southwest (also, the convenience of US 24-hour service and shopping).

My interview with Steen Rasmussen follows.

Suzan Mazur: Your interest in science comes from where?

Steen Rasmussen: My parents were just very supportive of these odd things I was interested in. Some of the early experiences I remember are my dad taking me out at night to look at the stars. We'd talk about infinity and about how far away the stars are. He'd tease me, asking whether I thought the universe would end and whether there was a wall. And if there was a wall, what was behind the wall. My mom would tell him to stop teasing me. So from the time I was a little kid sitting on my dad's shoulders looking at the stars, I've been interested in these questions.

Suzan Mazur: Also, Denmark is a wonderful science lab in the green, warm months.

Steen Rasmussen: Yes, when we have a good summer. The Danish summer can be gorgeous. There's light most of the night and it's warm enough to go swimming in the sea. We have many folk songs that praise the Danish summer. But the Danish summer is also said in folk songs to be this unfaithful beautiful woman who plays with you and then disappears, and you just get rain and cold weather after that.

You should be glad you're not here right now. The weather's terrible in the winter, grey, dark, wet and depressing. I wish it were snowing. I lived for 20 years in New Mexico and love it there. It's my other home, which I miss when I'm here.

Suzan Mazur: So why exactly did you leave New Mexico and the US and return to Denmark? And do you see a US - European rivalry regarding origin of life/protocell research?

Steen Rasmussen: Until recently, and since World War II, the US has invested significantly more than Europe has in all aspects of scientific research and development. That was why it was clear to me as a postdoc some 25 plus years ago that I needed to go to the US to develop my scientific career.

However, this US - European balance has shifted rather dramatically over the last 20 years, at least within my area. Today, I believe we have passed a tipping point where there is more basic funding available in Europe than in the US. Europe's young scientists don't have to leave for the US to pursue a career.

This shift was caused in part by the vision and implementation of new funding initiatives within the European Commission (Future Emergent Technology program) and the Danish National Science Foundation (Basic Science) and partly due to a significant decrease in basic research, *i.e.*, curiosity research funding across the board within the US. This was clearly felt at Los Alamos National Laboratory after the end of the Cold War. It was this shift in the availability of basic research funding that made me return to Europe. It was not because I wanted to leave the US.

Today I don't believe there is a will to spend public funds for origin of life research on either side of the Atlantic. However, we

270

do see cultural differences between science in the US and in Europe. Basic research is more an integral part of the scientific culture in Europe than in the US. In contrast, the utility aspect of what you do as a scientist is clearer in the US than in Europe.

In any event, it is very difficult to obtain funding to do blue sky origin of life research, *e.g.*, to develop protocells in either the US or Europe. However, scientists can address some of the origin of life questions within a context of bottom-up synthetic biology, self-replicating and repairing materials or living technology.

NASA as a funding agency in the US is an exception, as they actually have language in their research calls to support origin of life research. However, those funds are so minute that they are mostly symbolic in value, although I do see excellent work both by NASA postdocs and their astrobiology initiatives. NASA was created to fly and it is the science that supports space exploration that gets priority, not origin of life research.

I think the better strategy now is to seek origin of life funding from private US philanthropy rather than from NASA.

Suzan Mazur: You've been developing a protocell for about a decade. What is a protocell, that is, how do you define life, how close to making the protocell are you, and why are you doing it?

You made a statement about a decade ago:

> "By assembling one possible bridge between nonliving and living matter we hope to provide a brick in the ancient puzzle about who we are and from where we come."

Is this still your thinking?

Steen Rasmussen: Yes. It is.

Suzan Mazur: What is a protocell and how do you define life?

Steen Rasmussen: **A protocell is a physical-chemical implementation of the simplest life form that we can either make or that can emerge spontaneously.** I think there are many different ways we can make minimal living systems. Just to be clear, I'm speaking about the transition from nonliving to living materials, where you start out with components that are nonbiological.

They can be organic or inorganic materials or both. If you put them together in a particular way, you can create a system that can take in resources and convert those resources into building blocks for the system to grow and divide. Then if you have information, some kind of guidance for how this division process happens that is inheritable, and if the inheritable information can change from one generation to the next, then you have the possibility for selection. Because one kind of information control of how you grow and divide may turn out to be better than another information control of how you grow and divide, the better one will be selected and reproduce. As this process continues you have evolution, and then you're done. If you can implement a system that can do this, you have created a minimal cell.

The game, of course, is: How simplistically can that be done? A modern cell is a really complicated machine, but many of us think life can be created in much, much simpler forms.

Life for me is a physical process. In principle, living processes can be carried by different kinds of materials. You are not limited to biochemistry. In our laboratory, in experiments we did at both Los Alamos and here in Denmark, we were not using biological materials. We're using molecules that do not exist in modern biology. We're really building out of something that's different than what modern biology is using.

Again, in principle, you could build living systems out of robotics parts. We are not doing that, other research groups are pursuing that. But I think it's completely conceivable to have a macroscopic system that's able to build copies of itself.

Suzan Mazur: Why are you not working with biological material?

Steen Rasmussen: Life is a much more general process than what we see in modern biology. You can have living processes carried by robotic systems, by computational systems and by mixtures of biological, robotics and computational systems.

Suzan Mazur: I saw a reference to your definition of life and your colleagues' definition of life: The ability to evolve, self-reproduce, metabolize, adapt and die. Does that still hold?

Steen Rasmussen: **Yes. But to further address why we are not working with biological material, remember that modern biology has evolved over billions of years and has presumably developed sophisticated ways to solve the problems of being alive. I think there are simpler ways of doing that, maybe similar to the ways in which life emerged. We're trying to find such simpler ways.**

If you want to create life more simplistically than modern life is doing it, you can't use the sophisticated solutions modern life has evolved. It means you end up constructing your own building blocks. You have to build systems that are based on much simpler components. You actually need to have some of these components to carry more than one functionality.

For example, if you look at how the genetics in modern biology, the information and the metabolism interact with each other, it's a really complicated network of reactions and feedbacks.

To obtain the same functionality we have developed a very simple way, where the informational molecules, what corresponds to modern DNA, interact in a simple chemical way with the metabolism. There's a direct coupling so that electrons are jumping back and forth between the metabolic molecule and the informational molecule, which does not happen in modern life.

Suzan Mazur: How many labs worldwide would you say are now working on protocell development and which are the key labs?

Steen Rasmussen: It's difficult to say.

Suzan Mazur: I saw a figure of about 100.

273

Steen Rasmussen: I think there are more than 100, if the goal is working on different aspects of a protocell.

A few years ago, David Deamer from UC-Santa Cruz, the grand master of this kind of chemistry, went to the trouble of investigating how many labs there are, identifying nearly 100. I believe there are many more labs today due to increased interest in synthetic biology, artificial life and related fields. However, there are still not that many labs focused primarily on protocell development, if you look closely at their websites.

Suzan Mazur: You listed various research groups on your ProtoCell page:

> "Los Alamos Protocell Assembly; PACE (Programmable Artificial Cell Evolution); ECLT (European Center for Living Technology); Protolife; ECCell (Electronic Chemical Cell); MATCHIT (Matrix for Chemical IT); MICREAgents (Microscopic Chemically Reactive Electronic Agents); The Ribozyme Lipid Artificial Cell Initiative and COST-1; The Ribozyme Lipid Artificial Cell Initiative; and Szostak's--"

Steen Rasmussen: That website is no longer up to date. Many of these are projects I have been directly involved with or have some direct knowledge about. I'd say worldwide there are still less than 35 labs where they would say on their website: Yes, it's part of our job to develop a protocell.

Suzan Mazur: Is PACE still ongoing?

Steen Rasmussen: PACE is an example of one of these European-sponsored projects. We concluded that in 2008.

Suzan Mazur: PACE was working with a computing center in Barcelona?

Steen Rasmussen: Yes. That's right, Ricard Solé from Barcelona was part of PACE, but PACE was directed by John McCaskill in Germany.

Suzan Mazur: PACE's collaboration with the Barcelona computing center reminds me a little of what the origin of life group meeting at CERN may be proposing to do. You'll be part of the CERN meeting?

Steen Rasmussen: I was invited to the preliminary brainstorming meeting at CERN in 2011. . .

[**Note**: Participants for COOL EDGE 2013 had not been officially announced at the time of this interview, but Rasmussen was one of the presenters at the February 2013 meeting at CERN.]

It's wonderful that CERN is hosting the origin of life question. CERN is used to organizing large scale collaborative science teams, which solving the origin of life puzzle requires. . . .

How life originated on our planet is one way of thinking about this problem. But the problem really has two aspects. First, the historical aspect, for which we have no idea. We can't go back in time and see what happened. Second is the more scientific aspect, which has to do with properties of matter.

I'm much more interested in this second aspect. I'd like to know what it takes to turn matter from a nonliving state to a living state. This can be done in the lab through controlled experiments, by investigating all kinds of combinations.

Suzan Mazur: What are the main bottom-up approaches to building the protocell -- Is this the working list: (1) RNA (ribonucleic acid) world; (2) PNA (peptide nucleic acid) world; (3) self-reproducing lipid vesicles; (4) mineral surfaces-based metabolic processes; (5) cooperative feedback; (6) computational protocells; (7) Aromatic world - PAH (polycyclic aromatic hydrocarbon)?

Steen Rasmussen: You've mentioned a whole bunch of them. The bottom-up community seeks to build life from the bottom up, take building block by building block or aggregate by aggregate and put them together, so you can boot the system up and it takes off by itself.

Some bottom-up groups use the most suited starting materials they can find, they'll, *e.g.*, take building blocks from modern cells and try to assemble them in simpler ways. Other groups, including ours, are a little more minimalistic and will not take materials from modern biology. **Our group does not use enzymes. We don't use sophisticated lipids and sophisticated metabolic or modern DNA translation machineries.**

If you do work with starting materials that have evolved over four billion years and have sophisticated functionalities, you cannot know how these sophisticated functionalities have self-organized in the first place. You don't address the hard problem, how materials can transition from nonliving to living matter.

We want to understand how the nonliving materials, if you put them together in an appropriate manner, suddenly can become living. It has to do with understanding how matter organizes in a different way.

Think about the following. Look at your arm and then at the clothing you're wearing. If you look at the material the cloth is made of and the material your skin is made of, they are pretty much atom-to-atom, molecule-to-molecule very similar. But the properties of your clothing and the properties of your skin are very different.

If you rip your clothing, somebody has to sew it, but if you scratch your skin, it will grow together by itself. You skin has these marvelous properties and is organized in a very different way than the molecules in your clothing. We're trying to understand what it takes to organize materials so the system can self-repair, can grow and divide, replicate and evolve, adapt, utilize energy efficiently as well as have its entire set of components recycled.

There are fantastic properties we attribute to life. Such properties would be very, very useful for us in engineering if we could make technology that has some of the same properties. **That's why I think it's more fruitful to look at the scientific aspect of the origin of life problem, *i.e.*, how you can make nonliving material living rather than try to understand historically how** that happened, event for event, coincident by coincident.

If we can make artificial living materials, it would have huge implications for technology. We'll then be able to do things much smarter, much more energy efficient and much less resource consuming as all materials are recyclable.

For instance, when you and I die, all our materials can be recycled in the ecosystem. Having technology made up in a similar way for recycling would have great potential.

Understanding the scientific side of the origin of life problem opens these kinds of technological possibilities.

Suzan Mazur: Would you touch on microfluidics, which you've described as follows:

> "The life-cycle of the protocell is based on the self-assembly and division of lipid (fatty acid) micelles whose growth is driven by a simple photochemical process and controlled via genetic variability of informational peptide nucleic acid (PNA) replicators."

What is the promise of microfluidics for breakthroughs in medicine, including reading of PSA, and insulin levels, etc.?

Steen Rasmussen: Yes. It's certainly true that we are very interested in microfluidics and this interface between biotechnology and information technology. But it's not only microfluidics, it is a number of technologies connecting chemistry to information technology. To make it clear, I can maybe tell you a story?

Suzan Mazur: Yes, please.

Steen Rasmussen: This is how I think about it. If you look at science and technology in the last couple of hundred years, there was a major transition at the time of the Industrial Revolution. The essence of what happened is that we found out how to mass produce goods in factories, in an automated manner. At the same time, we built an extensive infrastructure to transport our resources and material goods.

The next really significant technological transition, which we are in the middle of right now, is the Information Technology Revolution. At the center of it is an automation of personal information processing and sharing in the personal computer and the Internet. It has given individuals the ability to access and produce and transmit information everywhere.

So the Industrial Revolution automated mass production of materials in factories and provided a mass transportation infrastructure, while the Information Technology Revolution automated personal information processing and an information access by the computer and the Internet.

Now look at living systems. What can they essentially do? They integrate material production and information processing in the most amazing way.

I believe that the next big technological revolution emerges when we merge information technology with material production.

To help us imagine how this material production and information processing could occur, let me first give a little background.

John von Neumann, famous as the inventor of the modern computer, is less known for inventing another machine called the Universal Constructor. Von Neumann proved in the 1950s that machines exist that are able to make everything including copies of themselves, as long as the construction process can be expressed as recipes (algorithms). This defines the Universal Constructor, which is a mathematical machine, which has not yet been implemented. However, it is a machine we often see in science fiction movies. You program the machine and out comes a tool, food or medicine.

We see some of the first primitive examples of such machine with the 3D printers.

Suzan Mazur: Can you say more about microfluidics used for medical readings -- insulin and PSA, for example?

Steen Rasmussen: One of the wonderful things about Information Technology is that you can program your computer. It's easy to give instructions. But it's very hard to tell a biological cell or biochemistry what to do. At the end of the day, however, all material objects have some chemical composition. So if you want to make new materials, you would need to control some chemical production. How can you instruct chemistry to do that? That's where microfluidics comes in.

You can program the microscopic flow of particular molecules in microfluidics by computers by actuators, *e.g.*, with electrodes or with other means. You are then able to control the chemical production down at the microscale, even down at the nanoscale. So you can make factories that are extremely small. We're still in the infancy of this technology, lab on a chip that can be used by individuals at home.

Suzan Mazur: But you're saying you already have this lab on a chip developed in some form that you can just plug into your computer?

Steen Rasmussen: Yes, but we can only do very simple things. What I'm referring to is the future where we'll be able to have Information Technology and biology or production technology to talk to each other, so you'll be able to program material production in the same way as you program your own computer on your desk today. You'll be able to have a personal fabricator able to make anything.

We'll eventually be able to implement von Neumann's Universal Constructor and make it into a Personal Fabricator you'll have on your desktop just as you have your computer and your printer today.

Suzan Mazur: Amazing.

Steen Rasmussen: We started to walk down this long path implementing simplistic versions of von Neumann's Universal Constructor a few years ago. The Future Emergent Technology

Office in the European Commission has already sponsored a portfolio of projects in this direction. **It is our technological vision that once we get to the point where we can combine biological systems at the microscale (bottom-up design) with what 3D printers can do (top-down design), we'll be able to have our own material production facilities at home. And then we'll be able to print our own medicine. Print our own clothes. Print our own electronics. We get to a situation where we are the designers, the producers and the user, just as we were before the Industrial Revolution. We'll have democratized material production.**

Suzan Mazur: I'm so glad you're working on it, Steen.

Steen Rasmussen: That's one of the technological derivatives of trying to understand the origins of life.

The intellectual part for me, what has kept me up at night, is this fascination with why materials in certain forms are alive and in other forms are not. Trying to figure out the secret of how to put materials together so they dance and become alive.

Suzan Mazur: Philosopher Jerry Fodor once said that our brains are not wired for the current rush of information. Is there a realization that humans as-is are just not going to make it very far into the future, even on this planet? Is that part of the reason you're developing the synthetic cell?

Steen Rasmussen: No, it's certainly not the reason why I'm developing the protocell. I'm developing the protocell because I have a deep fascination and awe for life and why we are here. I feel really lucky that it's part of my job to try to figure out how life came about and how we can use the wonderful processes of living systems to benefit civilization by making technology with some of these wonderful properties.

I agree, however, that it becomes a bit scary when you think about it on a geological time scale. **If you and I as humans create artificial living processes and machines that can copy themselves and ultimately evolve, where will this bring us? Considering that *Homo sapiens* is a very, very recent invention** in biological evolution, we as humans are naïve if we believe we are the end product of evolution. I'm sure we are not. So there's certainly something to think about.

Suzan Mazur: Are you saying you're concerned that we might actually create a Hal, that Stanley Kubrick's Hal emerges?

Steen Rasmussen: That's right. It's likely that we'll help create the next major evolutionary step. To help think about the negative possibilities of this development, we have all the Hollywood horror movie scenarios. Wonderfully described in *Terminator* **and** *Jurassic Park***. . . .**

[W]hat Hollywood describes is just one direction of a future living, intelligent technology. Another possibility is that we nurture our technology as we nurture our children.

We want our children to be happy, to be as smart, beautiful and insightful as possible as they grow up. We want our children to be able to go out in their world and do all the things they'd love to do.

Thinking along those lines, it might be possible that our technology will enable humans to create "things" that can make more beautiful poetry and art than we can, that can love more deeply and be more compassionate than we can. It doesn't need to be as depicted in Hollywood horror movies.

In any event, this next evolutionary transition does make me a bit uneasy, but I believe we have to move on. Dreaming, having curiosity and having the drive to invent new ways to do things better is one of the beautiful traits of humans.

Suzan Mazur: How far away is the making of a protocell?

Steen Rasmussen: A protocell, is of course, just a little step in that direction.

Suzan Mazur: But when do you think the protocell will happen? Other scientists are projecting within the decade.

Steen Rasmussen: When I look back at what I said about 10 years ago, I believed we might be able to do it in 10 years or so. But we're not there quite yet, and I've now been leading research teams on this for the last eight years. How much longer we'll have to wait for completion in part depends on how lucky we are with our next research grants.

Suzan Mazur: Freeman Dyson said "Give it another hundred years."

Steen Rasmussen: No, no, no… It won't take 100 years. There are a number of groups very close to having what is needed. Gerald Joyce a couple of years ago actually did make a self-reproducing and evolving RNA system. We've put together an information-controlled metabolic production of the protocell components. We still haven't got evolution going yet, but it certainly won't take another 10 years before we are done.

Suzan Mazur: Are you working with PNA?

Steen Rasmussen: Not anymore. It was too expensive. **The penalty I pay for trying to make this as simplistic as possible is that I have to chemically synthesize most of the molecules we use as building blocks. To build these molecules means that I have to employ synthetic chemists. And the more different materials you use, the more expensive it is. To synthesize new chemicals is both very difficult and time consuming.**

Suzan Mazur: Is there anyone working in the PNA world now?

Steen Rasmussen: Yes. Peter Nielsen was the inventor of PNA and the proposer of a PNA world. I still think PNA is a beautiful molecule. It's a very powerful one. It's just too expensive for us to use so we found a cheaper way.

My lab is combining some of the properties PNA has with DNA. We're modifying DNA, putting oily tails on it so the DNA can sit on the exterior of aggregates. We can thereby get some of the same functionality. I think we could have done it more elegantly with PNA, but it's just too expensive. Working with modified DNA is much cheaper.

Suzan Mazur: Didn't Dave Deamer comment that he thought PNA was a longshot because no one knew if it could reproduce?

Steen Rasmussen: I guess it has been demonstrated that PNA could be synthesized by prebiotic chemistry, but nobody has yet made PNA self-replicate. I think it is possible for PNA to replicate.

But I want to emphasize another problem in this connection. Chemistry is very, very difficult. It is difficult in a different manner than physics is difficult. **I'm a physicist, so one of the things I have realized when I put these research teams together is that chemistry is a bit of a black art. It's not like physics -- and it's not that physics is not difficult -- but the great chemists are like the great chefs. They have green fingers. You can have one chemist who can synthesize a particular molecule without any problem. Then you put the other chemist in and he uses exactly the same recipe but the soufflé falls flat. This is what is wonderful and terrible about chemistry. You simply don't know what you get before you try.**

Even though the protocellular systems we are working on are much, much simpler than modern cells, they are still so complicated chemically that it is very difficult to predict whether what we set out to do will work or not. We have to try it out. No theory or calculation can ensure that it will work or not. That makes it very difficult to make timelines and long-term plans, as some innocent-looking chemical step can become a roadblock for the whole project. And then you have to go back to re-try or find new ways around the roadblock.

So I'm more modest as a theorist today than I was 10 years ago. I'm more humble, because I've been living and breathing the problems that my chemists have had over the last decade. This is really difficult stuff. . . .

How do you get the first system to replicate? There's a process called self-assembly where you can aggregate materials and suddenly get new properties. For instance, if you have a lipid molecule in water, if you allow many, many lipid molecules to swim around in water, then they will form a membrane. Once you have a membrane, suddenly you can design an outside and an inside. You'll be able to define permeability through the membrane.

There are certain materials that can easily pass through the membrane and some materials that can't. It's logically impossible to observe these properties at the level of the individual lipid molecule. When you put things together you get new properties, genuinely new properties. This is a way to generate novelty, putting materials together in new ways.

So, for us to make the first protocell, we can't use evolution -- at least not evolution of the system itself. The way we have to get to a system that can undergo evolution is, we first need self-replication. We need to put materials together in a way such that the result of their interactions is that the whole system replicates. To make that happen you need free energy to drive the whole thing. That's why metabolism is also necessary as it can drive the replication process.

The individual building blocks of the protocell can't explain the higher-level structure and its functionalities when we put them together. They are an emergent structure. So we have to put a set of appropriate molecular aggregates together to get the first replicator.

This is the big scientific question -- figuring out which materials to put together to obtain a replicator.

It reverts to your original question about the origins of life. Once you have replication, then you can have evolution, but you don't have evolution before you have replication. At least I don't think so.

So when we talk about evolution, there are two important aspects to stress. We've already discussed the point regarding replication, but there's also the aspect of how innovative evolution is. Because depending on which evolutionary process you have, it can either be boring and only optimize existing solutions or it can be innovative in an open-ended manner and keep finding new solutions. Evolution is not just evolution . . . .

Norm Packard

NORMAN H. PACKARD
(*photo, courtesy N.H. Packard*)

I wanted to speak with chaos theorist Norman Packard for a couple of reasons. Packard was one of the principal scientists at Santa Fe Institute during its golden years in the 80s and 90s and has an understanding of the history of the science there. He also has a knowledge of synthetic cell development, as founder a decade ago of ProtoLife, a company originally headquartered in Italy focusing on "the bottom up approach of optimizing vesicles."

Packard establishes in our interview that protocell (synthetic cell) development was not really a big part of the discussion at SFI during those golden years, although autocatalytic sets was: "[T]here were various social and economic models that took insight from the autocatalytic network model."

The concept of autocatalytic sets (molecules catalyze each other's production) -- which theoretical biologist Stuart Kauffman considers his idea although admitting JBS Haldane may have been onto it first -- has been dismissed by Nobelist Jack Szostak as not realistic, at least chemically in protocell development. The following interview dips into that controversy and maps out synthetic cell development in general.

Norm Packard's PhD is in physics from the University of California, Santa Cruz. Aside from Packard's affiliation with SFI and ProtoLife, he co-founded Prediction Company with Doyne Farmer (another SFI scientist) in 1991 and served as its chief executive officer from 1997 to 2003, and chair until 2005. Packard and Farmer later sold the stockmarket predictive modeling enterprise to UBS.

In 2008, Packard moved ProtoLife to its present location in San Francisco. He says the company is working with "algorithms for optimization and discovery in a much broader context" and does not have an operational synthetic cell lab at the moment.

May 9, 2014 phone conversation

Suzan Mazur: How does the rethink on evolution, *i.e.*, that neo-Darwinism is dead, affect protocell research (*e.g.*, Margulis, Ayala, Denis Noble *et al.*)? That evolution is more top down and systemic -- tissues, cells, organs affecting the system – evolution's not gene-centered. The so-called genes are dead. How does this factor into current protocell development?

Norm Packard: My view of evolution is a little bit different from the view you cite. So maybe we can let that emerge. Evolution is a complex process that has causal effects both from bottom up and from top down.

Suzan Mazur: But the emerging thinking is that the so-called genes are dead, that they have to be affected in order for them to do anything. Some scientists are not even thinking in terms of genes as entities anymore.

Norm Packard: The assertion 'genes are dead' applies equally well to any particular material component of living systems. I like the observation generally, because it emphasizes that life is not about particular material components (which are dead), but is a process. The process involves many material components, among which are genes.

The theoretical story of what evolution is, is not at all cut and dried or universally accepted.

Suzan Mazur: I just interviewed Oxford physiologist Denis Noble, for instance. He edits the journal *Interface Focus* for the Royal Society. Noble's a leading systems biologist and is calling for neo-Darwinism (he calls it the modern synthesis) to be replaced. The late Lynn Margulis and Francisco Ayala have said neo-Darwinism's dead. "The Altenberg 16" a half dozen years ago recommended an evolution remix. Ever since the evo-devo conference Scott Gilbert organized over a decade ago, there's been growing evidence for the action being more top down -- tissues, cells, organs -- and systemic. It's not gene-centered, as was thought.

Norm Packard: It may not be gene-centered anymore, but that doesn't mean the genes are completely out of the picture. My perspective is that evolution is a complex process with causal effects coming from microscopic entities like genes, so genes definitely are players. You can call them dead if you want, but you cannot deny they are players in this complex chemical process that takes place in organisms.

This complex process of evolution also includes effects from the microscales due to other chemical processes. Very important are self-assembly processes that aren't usually taken into account in the telling of the story. There are also causal effects from larger mesoscopic and macroscopic structures like organs and tissues, and in general, the organismic context. The incomplete theoretical understanding has to include the microscopic as well as the macroscopic effects.

Suzan Mazur: Right. The thinking is that it's systemic.

Norm Packard: The role of evolution in protocells is -- and artificial cells are an interesting simple case because there we're trying precisely to make, to create a transition from chemicals interacting with each other both in terms of chemical reactions and in terms of self-assembly processes -- to create these mesoscopic entities that can start to be produced and evolve. So we're forced to understand what we mean by evolution in this very, very simple context.

To emphasize how simple this context is compared to the context of actual biology: In biology we have, even in the simplest organisms -- prokaryotic bacteria -- we still have complex structures with cells, with walls, and the cells are full of complex chemical processes, including chemical reactions that involve the genes of the cell to make all the proteins that the cell uses to survive and metabolize food and energy.

That context is already highly evolved compared to the context I'm referring to regarding protocells. The context I'm referring to has no proteins and no genes that encode amino acids to make proteins. It has no protein machinery. So what does evolution mean when you don't even have DNA encoding proteins?

Suzan Mazur: How many labs are working on the synthetic cell and would you name the labs? I'm aware of Jack Szostak's, for instance, and Vincent Noireaux's lab and his work with Albert Libchaber. Would you name some others?

Norm Packard: Before naming labs, let's explore the three approaches to creating protocells [synthetic cells]. There are basically three rather independent threads, each with its own distinct problems and its own distinct research pathways.

The first one to consider is the 'synthetic biology artificial cell,' most clearly exemplified by Craig Venter's path, which is to say, begin with existing living cells and simplify them by trying to reduce them to minimal genomes, then create an artificial cell [synthetic cell] by installing a genome into the living cell -- a genome you have complete control over.

Craig Venter is the paramount example in this case, although Venter's team is quite large and has other very famous people associated with it. I'm sure you know.

Suzan Mazur: Yes.

Norm Packard: With the Venter artificial cell, Venter *et al.* can install their own genome into it. They can put stuff into the genome that wasn't in the original genome. So, to some extent, that gives them arbitrary control over creating new cells by just installing new genomes.

Well that sounds great. But, of course, it's very difficult to figure out how to make new genomes so they are actually functional in new and interesting ways. And Venter *et al.* have made only limited progress in doing that. So that's the direction their research is going.

The problems they're facing are the problems that face synthetic biology generally, problems which arise because scientists are not usually installing entirely new genomes but just making modifications to existing genomes to create new functionality. It's extremely difficult to do that. Because when you make such modifications, typically you perturb the old functions of the cell, and it's very difficult to keep up the new functionality to get new cells to do something new in a really robust and production-level way.

That has happened a couple of times, but so far not with Craig Venter's artificial cell, to my knowledge.

Suzan Mazur: The second approach?

Norm Packard: The second artificial cell to consider is -- a completely independent path -- the bottom up artificial cell that's typified by Steen Rasmussen and his efforts.

What Steen wants to do is mix together the right components to make membrane-like entities and components to make metabolism-like entities and components to encode these other components and create robust self-producing, evolvable entities at the end of the day. This path does not begin with existing living cells. It just begins with chemicals.

Recently there's a third path to artificial [synthetic] cells, which is

290

to try and create cells by taking an empty cell membrane -- called a vesicle -- and then installing parts into this vesicle [minimal cell development]. When I say parts, I mean chemical processes basically. The chemical processes you install may actually steal from existing living cells.

They're not trying to take the entire genome of an existing living cell and install it in there. They're trying to install single, little parts, like for example a protein expression mechanism. You might also try and install an energy harvesting mechanism. An energy harvesting mechanism that you steal from some existing organism.

This pathway typically uses DNA and DNA chemistry and protein expression but in a way that Steen's bottom-up approach does not, but it doesn't use whole genomes.

More recently, the bottom-up approach has expanded to include efforts that combine chemistry with technology (electronic and microfluidic), to develop hybrid forms of living matter. An example is John McCaskill's recent projects, on which both Steen Rasmussen and I are collaborating.

Suzan Mazur: Which of these approaches are you using there at ProtoLife?

Norm Packard: The original work we did when we founded the company in Venice, Italy was on projects in collaboration with Steen Rasmussen, mainly on the European PACE project.

Suzan Mazur: ProtoLife is now based in California?

Norm Packard: That's correct. ProtoLife was involved in this bottom-up approach of optimizing vesicles for the protocell project. Subsequently our business has been based on using those algorithms for optimization and discovery in a much broader context. We're involved in lots of paths now. Most of our work currently is non-protocell.

We don't have a laboratory at the moment. We're a pure information technology company. We may get some contracts that will fund us to build up a laboratory here in San Francisco and we have partners that have laboratories. But right now we don't have a laboratory.

Suzan Mazur: What changed? Why did you leave Italy? What happened to the basic research funding that you had?

Norm Packard: The money went away in the sense that we were involved in this PACE project [Programmable Artificial Cell Evolution], which was funded for four years. At the end of four years the project ended, and the plan was always to bring the company here to San Francisco and launch the US operation on a larger scale.

The timing was not really felicitous for that plan because our move was in the fall of 2008 and the capital markets were not good. So we put ProtoLife's protocell development on the back burner.

Suzan Mazur: Who else is working on synthetic cells besides Craig Venter, Steen Rasmussen, Vincent Noireaux, Tetsuya Yomo, Jack Szostak?

Norm Packard: This third pathway or middle pathway that we might also call the biological parts pathway is an approach that Vincent Noireaux at the University of Minnesota is following. Noireaux has worked with Albert Libchaber at Rockefeller University in the past [and still collaborates with Libchaber]. Another extremely big player in this middle pathway is Tetsuya Yomo [University of Osaka]. There are some new players like Michael Jewett at Northwestern.

Then as far as the bottom up pathway goes, you mentioned microfluidics. Again, another important bottom-up pathway player is John McCaskill in Germany. McCaskill led the PACE project.

John is involved in this bottom-up approach more in a way of trying to create tight links between chemical and electronic and microfluidic technology.

Suzan Mazur: And where would you place Jack Szostak in those three approaches?

Norm Packard: I would place Szostak more in the bottom-up, though his lab uses several approaches, including the 'biological parts pathway.

Another example of bottom-up directions is Irene Chen, who comes out of Szostak's lab at Harvard. Chen is now at University of California, Santa Barbara. Then there's Martin Hanczyc at the University of Southern Denmark, who's working on oil drops as a platform for living processes. Another player who has got to be on your map with respect to this bottom-up approach, who's doing experiments trying to bridge the gap between nonliving and living with RNA-based chemistry is Jerry Joyce at Scripps.

Suzan Mazur: I met Jerry Joyce a couple of weeks ago at the Simons Foundation where he was lecturing on von Neumann's Universal Constructor machine. Joyce is now also a Simons Investigator on origins of life, he was awarded $2M by the Simons Foundation last year. John Sutherland and Matthew Powner, winners of Harry Lonsdale's origin of life prize, are also on the Simons Origins team with Joyce.

Norm Packard: Using whole genomes to create artificial cells the way Craig Venter does it is one example of the top-down approach. Basically I'd put most synthetic biology endeavors in this same category, even though they are not typically trying to make an artificial cell in the sense that they have a completely synthetic genome, but they are creating an artificial cell in the sense that they are creating new genomes that did not previously exist in nature.

They're creating new living entities but starting with living existing entities. I'd put all those efforts in the same general category as Venter's top-down approach.

Suzan Mazur: At the European Conference on Artificial Life last September, you and Roberto Serra *et al.* proposed a workshop on the protocell with a look to the past work to "glimpse the future." Has that workshop now been organized?

Norm Packard: I've just been approached by Roberto to be on the organizing committee of that.

Suzan Mazur: Any idea when the workshop might happen?

Norm Packard: It's a European complex systems conference, this fall -- September 22-26, 2014 in Lucca, Italy.

Suzan Mazur: Thank you. In Roberto Serra's review of the 2009 MIT *Protocells* book that you were one of the editors of, he was critical of the book saying the research was old. That some of the research was going back to 2003 and that most of it was no more recent than 2005. I was wondering if there was an updated version of that book in the works.

Also, Serra made the point that autocatalysis was somewhat underrepresented in the book. I wondered why that was.

Norm Packard: I don't disagree with his comments. There were size constraints. Given the material that we already had in there, we were faced with putting down the possible material that we could include simply to fit within the publisher's constraints. So that was one constraint.

Another constraint was that we were trying to include paths that had not only theoretical and modeling-based investigations but also experimental investigations. That's one of the reasons there isn't a whole lot on autocatalytic networks. Most of that work is in the domain of models and at the time lacked experimental realization.

Suzan Mazur: Is there a plan to update the *Protocells* book?

Norm Packard: Not at the moment.

Suzan Mazur: Is there anything political going on that you can speak to? For instance, I thought it was curious that Dave Deamer didn't have a single reference to Stu Kauffman in his 2011 book, *First Life*. And Jack Szostak told me he doesn't "believe" in autocatalytic sets, a concept Kauffman identifies with. Does it come down to belief systems?

Norm Packard: I think what you're experiencing here is tensions in a field where the scientific story is not yet fully understood. You have people coming from different directions with strongly-held opinions based on the directions they're coming from. They create scientific narratives they believe in -- yes they think they're right -- and the problem is these scientific narratives don't all fit together in a compatible, in a tight theoretical whole. You get these clashes when they're not fitting together. Typically the strongly-held opinions in these different quarters don't have good answers to the questions raised by alternative narratives because if they did, then they would have the tight scientific whole. That tight scientific whole does not exist.

I think some of the participants in this field behave as if it did. They behave as if their narrative is an unimpeachable scientific whole that doesn't have any significant challenges from the other narratives. I don't believe that's true.

I think Jack Szostak doesn't have this narrative. I don't think Stuart Kauffman has this narrative. I certainly don't have this narrative. But I know the different narratives well enough to say that the various conflicts among the narratives are not resolved to my satisfaction. The whole story does not yet exist.

I think scientists in this field would be making a healthy contribution if they were a little less dogmatic about their narratives and a little bit more open to discussing the conflicts among the narratives. There's not enough of that kind of exploration going on.

Suzan Mazur: One key European scientist told me he thought NASA's astrobiology mission was to control everything.

Norm Packard: Anybody can make a statement that institution X is trying to control everything on the basis of institution X has funding and controls the direction of research through the funding it's supporting. The EU has certain funding mechanisms and its scientific committees that determine its directions have certain inclinations. NASA has its direction and its funding, but the amount of money that NASA has dedicated to this is only one purse. I don't know what the level of funding is for origin of life.

Suzan Mazur: Five year grants of $40M total were awarded by NASA to five research teams in 2012 to study the origin and evolution of life in the universe. The money went to the University of Washington, MIT, University of Wisconsin, University of Illinois and the University of Southern California. There are 10 other NASA-sponsored research teams previously funded and working on the NAI roadmap.

One of the biggest centers is at Georgia Tech. Others are University of Hawaii, Arizona State University, Carnegie Institution of Washington, Rensselaer, Penn State, Ames, Goddard, and NASA's Jet Propulsion Lab.

About 1,000 people participate in the NASA astrobiology program.

Norm Packard: $40M may be a relatively big number to me and you but . . .

Suzan Mazur: [**Note**: In October 2014, NASA awarded another $50M in research funds for origin and evolution of life investigations. The University of Colorado received $7M from the pie to focus on "rock-powered life," life forms not powered by sunlight, but rather chemical synthesis.]

Private money is appearing. Harry Lonsdale, the Simons Foundation, Templeton, Jeffrey Epstein funding Martin Nowak's work at Harvard. That raises another point because the private sector doesn't have to be so transparent if they're doing the funding.

In one case, the project of one of the members of the secret peer review committee considering origin of life proposals was awarded research money.

Norm Packard: Yes. What can I say? Basically the problem you're pointing to regarding private funding is an issue with public funding as well. With public funding certain committees get formed to make decisions about what calls are going to be made for grant proposals and they then fashion the calls. And in the fashioning of those calls the committees can exclude certain research and include certain other kinds of research. It happens all the time. You have a committee with certain biases and certain directions.

But how can any committee not have certain biases and certain directions? They all have biases. So you end up with a process that's driven by certain biases. Origin of life research is a particularly vivid example of this because of two effects.

One is that opinions and biases about the origin of life happen to be very strong, which are the tensions you were referring to earlier. The second effect is the total amount of money funding this kind of research is very small.

So how can NASA be controlling everything, even if it has its biases, if its budget is so small? It feels like NASA is controlling everything because there's just not that much funding from other sources.

Suzan Mazur: I understand that autocatalysis is integrated into the European scientific system. Is there a divide in approaches to origin of life between Europe and the US?

Norm Packard: Yes, I think there is.

Suzan Mazur: And why do you think that is?

Norm Packard: As we've been discussing, different strong players with their own narratives, and their narratives tend to have certain attitudes toward autocatalysis and the role of autocatalysis in the origin of life.

Your book must face this issue at some point. There are issues with the RNA world approach. The main one is how do you get RNA starting to get produced in the first place. Right now, the way we see RNA getting produced is with this enzyme that catalyzes the production of RNA. But how do you get that enzyme?

Well that enzyme is a thick, giant protein. So somehow you have to bridge this gap of getting RNA production to happen without this enzyme or create a story of how the enzyme gets created along the way. But so far that hasn't happened.

Jerry Joyce has been doing some very interesting experiments [at The Scripps Research Institute] to get RNA to reproduce itself without that enzyme. Those experiments, by the way, start to look like autocatalytic networks of RNA interactions. Research on other forms of autocatalysis is going on in the US.

[**Note**: Theoretical biologist Stu Kauffman, who claims autocatalytic sets as his brainchild, has made similar comments re Jerry Joyce's work.

When I met Joyce at his Simons Foundation lecture this spring, he reminded me of the "tragedies" in Kauffman's life. Kauffman has publicized these tragedies in books, interviews, and most recently in a documentary he funded, which has been making the rounds of various film festivals.]

Suzan Mazur: I interviewed Nilesh Vaidya, the young chemist from Katmandu who worked on "autocatalytic sets" with Niles Lehman at Portland State University before leaving for Princeton and dropping the research.

Vaidya told me he and Lehman demonstrated that autocatalytic sets can emerge spontaneously. He said they put fragments of RNA in a buffer, added magnesium to water and the fragments then "stitched together" into an RNA enzyme. Vaidya claimed they replicated the experiment at least three times and got a different sequence of autocatalytic sets each time.

Nilesh Vaidya's name came up as a possible participant for the COOL EDGE 2013 origin of life meeting at CERN that Kauffman was "godfather" of, but Vaidya did not attend the Geneva gathering.

Norm Packard: Another example, a scientist who's been doing this quite a long time is Reza Ghadiri. He's developed autocatalytic reactions for peptides in much the same way as

Günter von Kiedrowski developed autocatalytic reactions for DNA.

But guess what? For some reason Reza Ghadiri hasn't made it a part of his intellectual world to really push autocatalysis of peptides in the origin of life story very hard.

Suzan Mazur: You think it's because of the politics?

Norm Packard: Ask him. He works down the hall from Jerry Joyce. And I'm sure he knows Jack Szostak.

Suzan Mazur: Vaidya is now working on unrelated experiments at Princeton. Niles Lehman is continuing the research at Portland State.

[Note 1: Lehman's research is partly funded by Harry Lonsdale's Origin of Life Challenge. Following are July 2014 comments from one of Lonsdale's three anonymous peer reviewers regarding the research of the team of Lehman, Higgs and Unrau, posted on the Lonsdale's web site (www.originlife.org):

> "Comments on 'cooperation-based perspective': I found this big picture overview difficult to understand. I'm not sure whether this is because the overview is poorly thought out or poorly explained. The key point is the autocatalytic set model, illustrated in Fig. 2b. While drawn so as to illustrate the authors' (or Kauffman's) vision of a closed autocatalytic set, I have trouble imaging how this would work given realistic sets of starting materials (the blue dots). . . . Given the lack of specificity of small templates, and the rarity of specific long sequences, I can't see how autocatalytic network catalysis can emerge in such a system. **I have to admit that in my opinion autocatalytic network theory has always been presented in a very abstract manner, perhaps to make it impossible to bring specific physical or chemical constraints to bear on the model. . . .**" (emphasis added)]

[Note 2: In my June 2014 interview with Nobelist Jack Szostak,

Szostak had this to say concerning autocatalytic sets:

> "Autocatalytic sets is one of those concepts where the people who came up with the original idea, like Stuart Kauffman, rather than admit being wrong, kept changing their story until it was basically the same concept everybody was already working on.
>
> The original idea was that there would be large numbers of compounds where one would help another to replicate, and that one would help some other one to replicate, and that somehow, out of this huge population of interacting molecules, autocatalytic replication would emerge.
>
> In my opinion, that was never chemically realistic. Now you see people talking about non-enzymatic RNA replication and calling that "autocatalytic sets." If that's what you want to call it, that's fine. But it seems like the concept has lost all meaning."]

Norm, how far back do you go in protocell [synthetic cell] development? Were you involved with it at the Santa Fe Institute?

Norm Packard: Protocells weren't really a big thing at the Santa Fe Institute for a long time. Generally complex adaptive systems (CAS) were the big story at SFI.

Stuart Kauffman had a variety of key complex system models that were part of SFI's intellectual and computational toolbox for complex adaptive systems. He had a whole library of them -- autocatalytic networks was one of them.

These models were mostly not involved in origin of life stories, but were used in other kinds of complex systems stories in the economics program at SFI. **There were various social and economic models that took insights from the autocatalytic network model**.

Suzan Mazur: I know Steen Rasmussen was involved in the early development of protocells, but I think he was doing it largely at Los Alamos.

300

Norm Packard: Yes, Steen was doing it at Los Alamos. There wasn't much on protocell development going on at SFI. Steen eventually came down and created some activity at SFI along these lines but most of Steen's activity was centered at Princeton, Los Alamos and then University of Southern Denmark.

Suzan Mazur: When did you first get involved with protocell development?

Norm Packard: With Steen and with John McCaskill in the creation of the PACE project in 2004, first fiddling with the ideas in 2002, 2003.

Suzan Mazur: In my recent interview with futurist Jaron Lanier, he voiced concern over the ramping up of artificial life/robotics research saying the following:

> "I don't even know if it can work in Europe. It can work in the early phase in Europe, perhaps, but you can't have a situation where you pretend that all the people aren't needed for anything and robots do the work. This gets to the illusion of Big Data. The truth is the only way to make machine learning algorithms work is robotics or autonomous systems. And it all depends on what we call Big Data, which means massive contributions from massive numbers of people. Without people creating examples, modifying them and reacting to them, the machines can't work. We're pretending that people are less needed.
>
> Now, in the early phase you can train people to work with the robots. If you adhere to the artificial intelligence ideology, then gradually you'll find a way to convince yourself that people aren't needed. But you can't have a 100-percent welfare state. So even Europe will break eventually. But the US will break first certainly, as we'll find out."

Would you comment?

Norm Packard: Yes, well that's some statement.

Suzan Mazur: What do you think about that? I was just looking at this conference that you were involved with in September 2013 in Sicily. It's staggering all the research going on with robotics and artificial life. What is going to happen to the human experience if this stuff really takes off?

Norm Packard: It's hard to say. I think that the narrative you just cited has some interesting perspectives but it's certainly not the final word by any means. The richness of the situation, the socio-economic system that we're participating in is itself a complex adaptive system in every sense of the word and it has basically very strong interactions that render a precise prediction of what's going to happen if you introduce some new element as unpredictable.

You can't really understand what's going to happen in advance, partly because you can't understand what the new elements are that are going to be created and even if you could, you couldn't figure out exactly what effect they would have because the effects are very non-linear and you don't have any way to make the prediction before you start making some observations.

So robotics and various forms of artificial life and various ways that Big Data can be used are all elements of what's happening in this complex system. That's why it's extremely difficult for me to have any faith in strong predictions because I don't think strong predictions are possible.

Part 5

Top-Down Tent

"The intellectual task is one of recovery, reclamation, and reminders of who and what we are and of what is being lost."
--**David F. Noble**, *Progress Without People*

Chapter 14

Send in the Robots?

November 1, 2013

Meandering through the cobblestone streets of old Bergamo on break from a recent conference, "Synthetic Modeling of Life and Cognition," I was drawn inside a shop displaying Pinocchio puppets of various sizes (the shopkeeper insisted the book's author was born in Bergamo, but in fact, he was Florentine). As the story goes, Geppetto the woodcarver, working from a piece of enchanted wood enabled Pinocchio – who had always dreamed of being a real boy -- to emerge.

Pinocchio 3000, the computer-animated film with Malcolm McDowell and Whoopi Goldberg has also done it, with Pinocchio the robot "brought to life by tapping into the city of Scamboville's power surge." But how close really are today's scientists to enabling a Hal or Pris & Roy to materialize?

Angelo Cangelosi, a 40ish year old researcher now at Plymouth University's Centre for Robotics and Neural Systems in the UK told me at a dinner party for the participants of the Bergamo conference that the European community has for the past seven years been funding cognitive robotics at €100M a year and that the EU will continue to support the research but that it is now making an even bigger financial investment in industrial robotics. As for a Hal or Pris & Roy -- says Cangelosi, "No, not in my lifetime."

Cangelosi's young colleagues gathered around agreed that robots will still have to somehow be plugged in, although not as obviously as Craig Ferguson's comedic sidekick Geoff Peterson.

Scientists I later spoke with at Santa Fe Institute (over gluten-free pizza) seemed to be keen on robotics. "Yes, robots are coming," said Nobel Laureate and SFI co-founder Murray Gell-Mann smiling.

David Orban, founder and director of Singularity's Institute for Artificial Intelligence - Europe, made a brief appearance at the University of Bergamo event and went further, insisting that people who don't embrace robotics in the future will not be able to survive.

Singularity co-founder Ray Kurzweil, now a director of engineering at Google, for instance, continues to say that building a human brain will happen by 2029: "[T]hat doesn't just mean logical intelligence. It means emotional intelligence, being funny, getting the joke, being sexy, being loving, understanding human emotion. . . . That is what separates computers and humans today." ("How Ray Kurzweil Will Help Make Google the Ultimate AI Brain," *Wired*).

But if life is non-algorithmic, how can a robot -- an algorithmic device -- become non-algorithmic? So, just where are we with developments?

I asked Vincent Müller, co-organizer of the Bergamo conference to weigh in. Müller divides his time between a position as research fellow at Oxford University's Programme on the Impacts of Future Technology and teaching philosophy at Anatolia College in Greece. Müller has written about the falsity of the mind -- he does not think humans have an exclusive on the mind -- and that building a robot with consciousness is possible.

Meanwhile, although there are now ways to enable brain cells to talk to one another as they lay on top of a bed of electrodes in a petri dish by pulsing them with an outside energy source, as cybernetics professor Slawek Nasuto *et al.* are doing at the University of Reading in the UK -- building a brain is another matter.

New York Medical College cell biologist Stuart Newman doubts that the complexity of brain cells and their collective interactions in humans, or even nonhumans, can be understood independently of their evolutionary histories in concert with their respective bodies. Newman emailed the following: "Robots, like other machines, are made of parts. Neurons are not parts in this sense, and a brain is no more a machine than the Roman Empire was."

Even if we succeed in building a human brain, how can we duplicate things like the delicate pas de deux of humans and their gut microbiota, for instance, which the International Society for Systems Biology conference in early September in Copenhagen made clear is crucial to human life?

Again, there is the matter of mind. We don't yet understand what a mind is or at what level consciousness begins, although theoretical biologist Stu Kauffman, in talks at Bergamo and in Sardinia, both of which I attended, said he thinks consciousness begins at the level of the electron and elementary particles.

Nonsense, said biochemist Pier Luigi Luisi to Kauffman's idea that electrons have consciousness. Luisi addressed origin of life at Bergamo and the week before at the Scuola Autopoietica del Mediterraneo conference in Sardinia advising that there's been nothing new on origin of life since Stanley Miller (Gell-Mann responded to the news by saying, "And I was there in Chicago when Miller did it."). However, Luisi and Kauffman both agree, that building a brain will not give you a mind.

The Italians and Japanese appear to be most enthused about trying to understand the role of "embodiment" regarding networks processing information. Osaka University's Minoru Asada who arrived in Bergamo wearing a white Borsalino hat, has been building a baby robot with biometric body (visual/auditory ability/tactile sensors) that simulates a child's development. His presentation ended with two robots flying into one another's arms like Anouk Aimée and Jean-Louis Trintignant (thanks to computer effects).

Vincent Müller's concern for the impact on society of robotics (pro and con), also as coordinator of the European Network for Cognitive Systems, Robotics and Interaction, has led to his organizing various forums and online chats about it. He thinks the robotics revolution will somehow translate into greater wealth for all, despite the fact that, according to an Oxford University report published this September, "The Future of Employment: How Susceptible Are Jobs To Computerisation," half of all jobs -- at least in the US are expected to disappear in the next decade or two due to computerization. The report did not mention that the cost of job retraining in the US will be left to the unemployed to shoulder.

Slawek Nasuto --

Darling, Which of Us Is the Robot?

SLAWEK NASUTO
(*photo, courtesy S.J. Nasuto*)

"We see the world in terms of trees and dogs and rivers, etc. But often the question is, what are those concepts?" -- **Noam Chomsky**

December 2, 2013

Cyberneticist Slawomir Jaroslaw Nasuto has a radiance that can only be Polish. He was born in eastern Poland, in Lublin, during the Soviet years (Stalin was embraced by the post-WWII provisional Polish government headquartered there) and is now doing ground-breaking research on the brain and embodiment at the University of Reading in the UK. Some of this work he shared weeks ago at a cognitive robotics conference in Bergamo, Italy --

where we met -- in particular, his experiment with the "animat": a CLOSED-LOOP system of fetal brain cells in a dish over electrodes + computer + robot.

Slawek (pronounced "Swah-vé") Nasuto was fascinated by math and science at an early age, as well as by Western literature smuggled into Poland. His father was a professor of physical chemistry at Maria Sklodowska-Curie in Lublin, the university named for Madame Curie, the first woman to win a Nobel Prize.

Nasuto's MS degree is in pure mathematics from Maria Sklodowska-Curie and his PhD in cybernetics from the University of Reading. He is currently associate director at Reading's Centre for Integrative Neuroscience and Neurodynamics, and he led the initiative to establish the new Brain and Embodiment Laboratory (BEL) there at the School of Systems Engineering, where he serves as a director.

BEL will analyze data at multiple scales, including multi-channel electrophysiology and EEG in various closed-loop experimental paradigms.

Nasuto has authored and/or co-authored 120 papers in computational neuroanatomy and neuroscience, the role of synchronization in cognitive processing EEG-based brain - computer interfaces, the relationship between structure and function in individual neurons and their networks.

Despite some of the challenging questions I put to Nasuto during our conversation, he remained upbeat about the benefits from robotics research. His enthusiasm and that of his colleagues at the University of Reading has also caught the attention of BBC.

I last saw Slawek Nasuto in September in the grip of one of the alpha males at the Bergamo event who was trying to extract from him just how origin of life might intersect with robotics -- the answer so far remains elusive. . .

My interview with Slawek Nasuto follows.

Suzan Mazur: What was it like growing up in Poland during the

Soviet years?

Slawek Nasuto: Not so remarkable. I was born in Lublin, a city in eastern Poland. I spent all my childhood there. I had a very influential math teacher in secondary school, a really strong character. Thanks to him the majority of our class -- 17 - 20 students out of 30 -- went on to study mathematics. Quite unusual. I was actually interested in science from a young age.

Suzan Mazur: Do you come from a science family?

Slawek Nasuto: Yes, my father, now retired, was professor of physical chemistry.

Suzan Mazur: Where was he a professor?

Slawek Nasuto: At the University of Maria Sklodowska-Curie in Lublin.

Suzan Mazur: Was your mother also a scientist?

Slawek Nasuto: She was not. She has always been an avid reader. I developed a passion for reading from my mother.

Suzan Mazur: What else stands out about those years?

Slawek Nasuto: Poland was relatively isolated from the Western world and I was not very much interested in politics. Kids tend to think of the environment they grow up in as normal. If a kid doesn't have an example of an alternative, it's difficult for them to compare.

Suzan Mazur: The Soviets were promoting science.

Slawek Nasuto: Right, yes. Science books in Poland were very cheap. Translations were actually mostly from the East Bloc. For anybody interested in culture and reading it was also possible to find books.

When I was growing up the censorship wasn't that strong, so many of the books, even Western classics were available. Some were not. As a teenager I remember reading George Orwell's

Nineteen Eighty-Four. My best friend got the book somehow. It was an illegal copy smuggled to Poland from a publishing house in Paris and translated into Polish.

Suzan Mazur: So you benefited from an East and West perspective in science?

Slawek Nasuto: To some extent yes. In secondary school I was very, very interested in popular science.

Suzan Mazur: What are some of your interests outside science?

Slawek Nasuto: I tried a lot of things when I was younger. I was extremely skinny as a kid so I began working out at the gym. My parents were afraid that I had to put some weight on, to develop muscles or I'd have problems with posture. I've been going to the gym ever since. Also, I have a fascination with the Far East and have investigated different martial arts. I like sailing as well.

Suzan Mazur: Did you say your family name Nasuto may be Eastern?

Slawek Nasuto: We don't really know where it comes from. It's a very rare name in Poland.

Suzan Mazur: You led the University of Reading's ground-breaking research on "animats." In your recent papers, you describe the animat as a system with culture and robot coupled via a closed loop, the culture consisting of tens of thousands of dissociated cortical neurons and glia cells taken from a fetal rat and placed in a petri-like dish with a grid of electrodes embedded at its bottom.

I understand the culture was then pulsed with electrical signals from a computer (also part of the closed loop) and within an hour of placement there was a reconnecting with other neurons and internal communication -- chemically and electrically -- even without the electrical pulsing.

Slawek Nasuto: Actually, all this happens relatively gradually with the first signs of cells reconnecting within hours from seeding and then activity coming on once the connection reaches a sufficient density. We have found considerable variation between cultures, as well, because there are experimental factors we can't precisely control, *e.g.*, seeding density of cells, and also due to the intrinsic variability of cultures. So there are large 'error bars'.

Suzan Mazur: The robot (part of the closed loop) was wirelessly controlled using bluetooth, a system also used to print photos from a mobile phone, for instance. Can you round this out for us? How soon after placement was the culture signaling to the robot?

Slawek Nasuto: The culture typically takes some time for the cells to settle. Once the cells settle, they start to form connections. As the culture becomes more dense, it begins to communicate -- typically we see some level of connectivity and activation at 24 - 48 hours.

One goal of our research was to connect the robot to the culture as soon as possible and let it develop as it interacts with the environment. We could not achieve that in the original project for technical reasons. We are about to restart the wet lab in the university's Brain and Embodiment Laboratory so this line of research is going back on the agenda.

Suzan Mazur: You say the culture matures at one month and lasts for three months. Meaning the animat "dies" after three months?

Slawek Nasuto: Right "dies" with quotes, the culture is not a sentient being. In some of my papers I argue this is actually not the case (*e.g.*, the zombie mouse). The culture will survive as long as laboratory conditions enable it. It's possible to keep cultures for even longer, given sufficiently stringent laboratory procedures, but it is tricky. What's important is that we were able to keep the culture alive for up to three months.

Suzan Mazur: Do you toss or preserve the experiment when it dies?

Slawek Nasuto: We make records of the culture. When it dies, it is disposed of.

Suzan Mazur: Did you also experiment with adult brain cells, and, if so, did they exhibit a similar plasticity?

Slawek Nasuto: We used an immature culture and waited until it matured before we connected it with the robot. Again, one of the goals was to connect as early as possible. We did not start any of the experiments with adult brain cells. Fetal cells are used because they're easier to maintain in a healthy and responsive mode. But, yes, even in adult cells there is some plasticity.

Suzan Mazur: You conclude in your paper "Controlling a mobile robot with a biological brain" that "a robot can have a biological brain to make decisions," that there's a need for further investigation. But do you envision those robotic decisions in the future to be meaningful decisions?

Slawek Nasuto: That's a million dollar question. Let me answer indirectly. When we were building the animats, we didn't think about whether the robotic decisions would be meaningful, in the intuitive understanding of the term. This was a ground-breaking experiment, our very first approach.

We treated the culture as almost an input-output device. That by itself is not really taking too much of the culture's biology into account. But when we talk about systems supporting cognitive processing, we are dealing with a system, the brain, structured by evolution and development. We have evolved very special signaling pathways and different centers in the brain thought to be involved in some cognitive processes.

If we take this biology seriously into account in construction of future animats, I'd say the jury is out as to whether robotic decisions can be meaningful decisions.

Suzan Mazur: What is the big lesson learned from this experiment?

Slawek Nasuto: In experiments in which one uses tissue from living beings, the objectives tend to be much more specific, questions focused on tangible (even if ambitious) aims that can more directly counterbalance the ethical issues inevitably associated with experimentation.

Big philosophical questions may or may not be answered but they by themselves cannot be the only reasons to conduct such experiments. One of our long-term goals was to create an alternative experimental platform for testing novel pharmacological agents to treat neurological disorders. For that we needed a system that offered us great access to information about how the animat works.

Its operation may be plausibly mapped onto information processing going on in the brain/nervous system when it engages in cognitive processing. At the same time the animat is not a sentient being, so that it does not suffer because of the experimentation.

I believe the animat will eventually be such a platform. In the future the creation of an animat capable of advanced information processing will contribute to a reduction of animal suffering in experimentation. This will also offer us unparalleled access to data as the animat processes information -- data we cannot for technical and ethical reasons now collect so easily (or at all) from animal studies.

It was in our Templeton-funded project that we tackled the question as to whether such animat could be thought of as a sentient being. We concluded that it could not. However to answer this we needed to look at the meaning of computing, Turing or otherwise. We argued that arbitrary formal manipulation of nervous tissue does not lead to any cognitive capacity in it.

This is quite important because, if animat, which after all has a biological "brain" to speak of, can not be sentient, so can not robotic devices. No matter how sophisticated or human-like in appearance they might be.

Suzan Mazur: That leads to my next question. You cite in the above-mentioned paper, which appeared in *Defence Science Journal*, a publication of the Indian government's defense department, the experiments of others who successfully sent control commands from a lamprey to a robot, and also from a mobile phone or PC to living animals, such as cockroaches and rats, with electrical implants and microprocessor backpacks. You question the ethics of such experiments. Here's the quote:

> "Regehr demonstrated that it was possible to use the brain of a lamprey to control the trajectory of a robot whilst others were successfully able to send control commands to the nervous system of cockroaches or rats as if they were robots. Although such studies can inform us about information processing and encoding in the brains of living animals, they do pose ethical questions. . ."
> -- Warwick, Xydas, Nasuto, *et al.*, "Controlling Mobile Robot with a Biological Brain," *Defence Science Journal*

Would you comment further? What are the implications? Where could this lead us?

Slawek Nasuto: These studies are trying to understand how the nervous system works in order to understand when it goes wrong. So we can create better therapies and interventions, etc.

The problem in the case of the experiments you just cited, obviously, is that they are performed on sentient beings. The tests are carried out on a nervous system or on an entire animal. We can argue to what extent these animals have consciousness, and what that level of consciousness is, but they do have feeling. Hence the ethical questions become important when such experiments are performed.

Suzan Mazur: Well why is the Indian government, the Indian Ministry of Defense interested in this?

Slawek Nasuto: I think it was just an open journal. I don't think there was any specific interest from the Indian military.

Suzan Mazur: Is there any understanding how far this can be taken at present in terms of controlling the nervous system of other animals and people? What's doable?

Slawek Nasuto: At present, there are different kinds of research on hybrid systems, technologies interfacing with the nervous system. Ted Berger in the US, *e.g.*, in experiments with rats, and more recently, monkeys, is attempting to build an electronics system able to restore memory formation. The system is based on a chip implant that collects data from one area of the hippocampus and applies an appropriate transfer function before passing the activation on to other areas. It was demonstrated to mimic the operation of a damaged natural information pathway. This work can be very promising if it can scale to humans and to more natural scenarios than lab-constrained experiments.

My colleague from the University of Reading, Evangelos Delivoupoulos, has also been involved in development of flexible electrodes. Electrodes are typically placed either in the sensory cortices and help decode neural signals, or in the motor cortex in order to learn how the brain controls movement.

The flexible electrodes Delivoupoulos has been developing can be chronically implanted in nerves controlling the bladder -- technology that offers promise in increasing quality of life, autonomy and dignity to people with lower spinal cord damage.

Suzan Mazur: What are your thoughts about the possibility of a non-algorithmic Trans-Turing System -- "a new class of information processing system" -- and the statement that one approach would be to "Simulate the TTS on a digital computer and evolve a population of TTS" in light of what Oxford philosopher Vincent Müller recently told me, which is:

> "My view is anything to do with our computer systems is fundamentally algorithmic. If the system that we're trying to generate on that is not fundamentally algorithmic, then we won't be able to generate it."

Slawek Nasuto: The answer is not straightforward. Personally, I agree with Vincent Müller's statement, although the problem is that during the course of research in this area I came to appreciate as well that even in computer science there is no universal agreement on what computers can and cannot do. I'm trying to provide an answer that does not cut corners in my reply because this is one of the cases where I believe simplification leads to further confusion.

The answer is not simple, in part because we use computers every day in science and engineering to approximate functions that are inherently non-symbolic. This is the pragmatic part of the answer. It does not address in a fundamental way the question of Turing computing limitations of computers but it blunts the strength of the fundamental objections to their computing power.

The more subtle answer is that we have also begun to use computers in ways that go beyond the reach of formal algorithmic descriptions. Operating systems, word processors, Internet. These are just a few examples of such uses.

One of the characteristics of the above examples is the inherent interactions with the outside, the environment. The purpose is to engage in a successful interaction rather than compute specific function.

Such systems go beyond the computational capacity of a classical Turing machine, but that does not mean there is anything mysterious about them. For sure, they are built of algorithms, so each well-defined fragment is subject to the formal Turing-style description. But they are put together in a way that escapes such a formalism.

To give an analogy, it is a bit like a link and a chain. A link made of metal is rigid, cannot bend. Yet a chain made simply of such links sequentially interlocked can be as flexible as a rope. The system acquired a characteristic that is not present in the individual element.

What I am saying is that even with the Turing machine we can do things that go beyond Turing computability limitations. Because of computers we put a man on the moon. Tremendous advancements in science have been made thanks to computers that can approximate.

But the notion of computing is subject to active research, and extensions of classical computing have been proposed. In our

317

group at the University of Reading we are trying to look at the possibilities of clarifying these issues. What computing is. Because there is a lingering belief in some areas of cognitive science that the brain can be thought of as performing formal manipulations on symbolic representations of reality akin to a Turing machine. And the Turing machine is a very specific model of an algorithmic process.

Our Reading group, in its investigation of what is computing, started from the premise that the notion of computing -- in spite of its usefulness -- cannot fundamentally describe mind or cognition. Computing thought is an inappropriate statement.

Suzan Mazur: The Turing Machine was never actually built.

Slawek Nasuto: The Turing Machine is an abstract concept [never actually built], a formalization of a mechanical process. But the process was well defined by Turing. Again, computers that have since been constructed we are using in ways that cannot merely be described as algorithms because the computer systems are engaging, interconnecting.

Suzan Mazur: Will you be returning to Poland anytime soon?

Slawek Nasuto: Never say never. But the Brain and Embodiment Laboratory here at the University of Reading is almost fully functional. It took two or three years to convince the university to give us the lab and I'm very much looking forward to some fascinating long-term projects.

Vincent Müller

Robots and the 1%

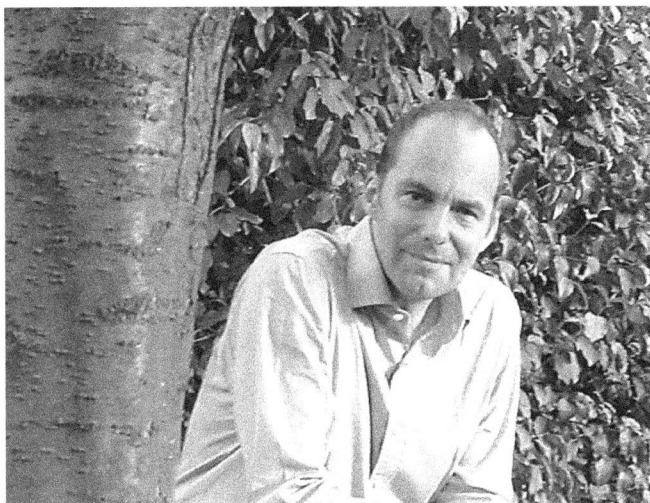

VINCENT MÜLLER
(photo, courtesy V.C. Müller)

November 11, 2013

Philosopher Vincent Müller has been to the mount. To the peak, of Mt. Olympus that is, many times, as a mountaineering enthusiast who now calls Greece home. These days Müller thinks a lot about how artificial intelligence will -- pro and con -- impact humanity, both in his role as a teacher of philosophy at Anatolia College in Thessaloniki (an hour's drive from Olympus) and as James Martin Research Fellow at Oxford University in the UK.

But it will take an infinite number of conferences in the palace of Zeus before the answer to that question is clear. For now, the road ahead is a minefield with huge numbers of unemployed people worldwide not keen on being permanently replaced by robotics and aware that the so-called social momentum, the madness for robots -- at least in the US -- is driven by the PR machine of the wealthiest 1%. In other words, without consulting Zeus, the 99% can intuit just who will be getting richer.

I met Vincent Müller in September at a conference in Bergamo he co-organized on Synthetic Modeling of Life and Cognition. Following the conference we had a chance to talk informally over dinner at a lakeside restaurant outside the city.

It was Yom Kippur, and Müller (German) and another of the conference presenters (a Jew, with a masterful German accent) after a glass or two of local wine began trading Jewish - German jokes across the table. . . .

Müller has a somewhat furious schedule these days, including a commute every other week between Thessaloniki and Oxford. He is coordinator of the EU's European Network for Cognitive Systems, Robotics and Interaction and somehow is finding time to write a book on the problems of artificial intelligence as well as edit several volumes on the theory of cognitive systems and AI.

Vincent Müller's BA, MA and PhD are all in philosophy from, respectively, Phillips University Marburg, King's College London, and Hamburg University with postdoc work at Oxford and Princeton. I spoke with him by phone recently at his home in Greece. Our conversation follows.

Suzan Mazur: Can we begin with some background? You divide your time between Oxford University where you are James Martin Research Fellow looking at artificial intelligence and a professor of philosophy at the American College of Thessaloniki/Anatolia College in Greece, plus you are coordinator for the EU's European Network for Cognitive Systems, Robotics and Interaction. Plus you are working on a few books on artificial intelligence and have organized several conferences on it. You note in your biography that you've organized a mountaineering club. Let's start there. Tell me about the mountaineering club, if you would.

Vincent Müller: I've been mountaineering since early childhood, first with my parents and my grandparents. After I moved to Greece I discovered that it is essentially very mountainous, very beautiful mountains, even though most people think of Greece as a place of sunny beaches. I began organizing trips with my students to some of these surrounding mountains, including Mt. Olympus, the highest point in Greece as well as a place of great cultural significance. I haven't participated in many of these trips in the last couple of years because of my professional work and enjoying time with my family, but I do very much like mountaineering when I have the time. I was on Mt. Olympus just two weeks ago.

Suzan Mazur: You were you born in Germany, weren't you?

Vincent Müller: Yes, I was born in and grew up in Germany. I studied there as an undergraduate also. Then I went to the UK, to King's and Oxford, and returned to finish my PhD in Hamburg, followed by postdoc work at Princeton. I later got a job in Greece and moved here.

Suzan Mazur: Would you give me an idea of just how large the artificial intelligence and cognitive systems community is worldwide in terms of researchers?

Vincent Müller: That's not easy to say, especially because the borders are not clearly drawn. We have around 900 researchers (a PhD student or above) in our European EUCog network. Perhaps that is about 20% of the total number in Europe, so say 5,000 in Europe and 25,000 worldwide?

Suzan Mazur: That's quite a lot of researchers.

Vincent Müller: The European Commission is spending €80M per year on the field, but there must be a lot more going on. Europe is one of the largest single funding bodies worldwide -- €80M is quite a lot, particularly when compared to the funding of philosophy.

Suzan Mazur: In an upcoming book you're looking at some of the problems of artificial intelligence. Would you identify a couple of the problems? What percentage of the population is opposed to the ramping up of automation, for instance?

Vincent Müller: The "problems" I discuss are primarily the ones we need to overcome to achieve a higher level of success with these devices. It is very clear that artificial intelligence has the potential for significant impact on humanity in a positive and negative sense and that is something I work on at Oxford. AI already has a very significant impact on the future of the self image of humanity, of how we see ourselves.

Suzan Mazur: The late science and technology historian, David F. Noble said the following almost 30 years ago in an article for *Science for the People*, which Noble later developed into a book: *Progress Without People*. Here's what he wrote:

> "[I]t is the support of those in power (in our society, those with money, or those with political, military, or legal authority) that affords technical people the luxury to dream, to dream expansively (yet within) well understood limits, and to make their dreams come true (by imposing them on others)."

Would you comment?

Vincent Müller: It's important to see that scientific and technological research are social phenomena, bound by the social and power structures within which they take place. It's doubtful that this research is a manifestation just of the powerful plus the dreams of scientists. The kind of research that we carry out is pushed by significant social support of it, both politically-guided support -- publicly-funded areas -- and economically-guided support, directed by common interests.

Suzan Mazur: That may be true in Europe but in the US the public does not have a say over what is done with publicly-funded research. This is a point David Noble made clear in our interview several years ago, that decades ago citizens were cut out of the decision-making process:

> "**David Noble**: The Vannevar Bush *et al.* legislation said

322

essentially that science would be funded by the taxpayer but controlled by scientists. Again, scientists -- this is important to emphasize -- are not simply scientists, but scientists and the corporations they work for . . .

There was a problem with the way the committees and panels overseeing the allocation of research funds would be set up. The problem had a name and the name is DEMOCRACY. The fundamental tenet of the democratic system is that the taxpayers funding something have control over what's done with the money.

Harry Truman said it was the most undemocratic piece of legislation he'd ever seen and vetoed it. It went through minor changes and became what we have today -- a scientific establishment run by scientists with very little political oversight. The key thing is how they kept the taxpayer out was through peer review."

Tony Prescott's recent report in the CLAWAR eJournal mentions that a Eurobarometer survey found 60% of people polled in Europe think robots should be banned from use for children, the elderly and disabled citing possible psychological damage.

Vincent Müller: Yes, but people see this very differently if you ask them for technical means that enable people to be more autonomous.

Having said that, whatever the social power structure is that leads to research funding taking this road or that, it is clear that a very significant amount of research is beneficial to most of the population because it results in economic productivity. Now we have much more powerful computer systems than we had 15 years ago. That enables us to be more productive and generate more wealth.

Suzan Mazur: The EU-funded "factories of the future," initiative was designed to make industry in Europe more robust through information technology that would ultimately lead to more jobs. Is it unfolding as designed? Are those jobs materializing?

Vincent Müller: That's a very complicated question. It's something that's being debated at the moment, whether robotic automation is just an aspect of "normal" technological development and improved efficiency. If that is the case, then a technology produces some technological advantage, generates a higher efficiency in one area of production, and thus reduces jobs, but at the same time produces employment for others. Overall, society benefits. There's more prosperity so the unemployed are able to find new jobs. There's more wealth generated.

Suzan Mazur: But it's the 1% who increasingly have the wealth.

Vincent Müller: When I was a child there used to be typesetters, people who set little lead pieces in rows for printing. When electronic typesetting and later desktop publishing machinery arrived these people lost their jobs very quickly. But that meant that newspapers and other printed matter could now be produced much more quickly and cheaply. So overall this was an economically beneficial development. Some people had to be retrained.

In a social democratic society there would be a state program to retrain so people don't just fall by the wayside. The question is whether this is a development we can continue indefinitely or whether there is some kind of limit. Can we keep replacing people's jobs with more sophisticated machinery?

Suzan Mazur: The US is not a social democracy, however, and the cost of retraining here will be left to the unemployed to shoulder.

A report by your colleagues at Oxford on the future of employment indicates that the reverse will happen in the US regarding creation of jobs. Carl Benedikt Frey and Michael A. Osborne in their recent study say that in the next decade or two roughly half of all jobs in the US will be gone due to technology. If you want a job, go to beauty school.

Some analysts here say this has already happened. The Oxford report also cites McKinsey's findings that 44% of all firms that downsized since 2008 did so through automation.

And try submitting a resume to the Human Resources Department of US companies without a knowledge of the computer catchphrases. The question is, how will people survive in a world like this of robotics?

Vincent Müller: I don't want to comment on the technical details of Carl Frey's and Michael Osbourne's report. It's an economist's question and I'm not really qualified to do that. However, they are trying to gauge how many kinds of jobs could potentially be replaced, not how many people will be replaced. The replacement of humans by machines has been happening for a very long time, basically throughout the industrial revolution.

Suzan Mazur: Yes, but companies in the past trained people and paid their trainees. Now in the US the norm is an unpaid internship -- MAYBE -- following training paid for by the intern. How do people survive if they don't have jobs or money to get retrained? Is the point to downsize the population?

Vincent Müller: If something of that sort is actually going to happen, and we don't know yet whether that's true – it's not happening yet. But if something like that is going to happen, then we would obviously need to find a way to negotiate this very significant change as we did in the past when jobs in agriculture were almost wiped out.

Suzan Mazur: Are you saying that we should take a wait and see approach to see if people are beginning to die off?

Vincent Müller: As an example there was an enormous uprooting of entire populations as agriculture went from manual to machine-driven labor. Before World War II, the vast number of people were working in agriculture. Now in most developed countries less than 10% do so. So the question is whether (a) we will see a development like that, and (b) whether we will be able to turn that development into a positive for the society with more wealth generated for all or whether we will reach a point which economic historians have been talking about for a long time where we will not have enough work.

Suzan Mazur: Frey and Osbourne point out that the US high school movement in the early part of the 20th century eventually took us to a point where there was a "supply of educated workers outpacing the demand for their skills." This situation is now acute. They say "high-skilled workers have moved down the occupational ladder, taking on jobs traditionally performed by low-skilled workers, pushing low-skilled workers even further down the occupational ladder and, to some extent, even out of the labour force." That's the situation now and it continues to get worse. Very difficult.

Vincent Müller: If it's a matter of improving efficiency, which industry has been working on since it started, then I don't think it's a difficult problem. If we're looking at a categorically new thing, in which the movement of labor will not happen, then we are facing a substantially more serious problem. But it's a problem we'll be facing in the context of substantially increased wealth. And in that context we'll be able to deal with the problem.

Suzan Mazur: Again, the wealth is in the hands of the 1%.

Vincent Müller: The issue is whether we can make that wealth for a large part of society.

Suzan Mazur: As a philosopher, haven't you argued for the falsity of humans having minds and said that "the mind is dead . . . forget the mind"? Would you expand on that and put those statements in context?

Vincent Müller: Quite a different issue. When I said the mind is dead, what I meant was that I find this a very un-useful concept for explaining what's going on, which shows in the context of the 'extended mind' debate. For example, in the context of free will, I think there is a useful description of the cognitive structure that allows strong beings to have that feature, and there is no particular reason to think that human beings have an exclusivity on that feature, in principle. And there's also no reason to think that having the feature of free will requires that the world is somehow not deterministic. "Having a mind" only muddies the waters here.

Suzan Mazur: Do you make a distinction between conscious and cognitive?

Vincent Müller: Yes. There are different procedures or processes that I would call cognitive and some of these are conscious and some of these are not.

Suzan Mazur: I'm still unclear about your statement that the mind is dead . . . forget the mind.

Vincent Müller: What I mean by that is that you cannot base the whole concept, to call something mental or not mental or the mind and not the mind in order to understand what is going on for humans or others. It's been an unfortunate term -- "mind" -- in the history of understanding cognitive phenomena. Many languages don't have this term "mind."

My reference to the mind was in relation to the extended mind. And the discussion of the extended mind boils down to a borderline question -- what is mental and what isn't. That's the part I think is unhelpful. But that's not to say that all the terms that we've been using in the description of the mind are unhelpful.

So for example, the word consciousness has been used in several ways that are useful. There is consciousness in the sense of having experience of what is life, known as phenomenal consciousness. That I think is a useful term.

Indeed, if we think that that is what we mean by consciousness, then there is an interesting point as to whether we would have the ability to generate artificial beings with that feature. It's a difficult question because we already know there's no particular scientific test by which we can find out whether a creature other than ourselves actually has phenomenal consciousness. This is known as the other mind problem.

Consciousness is one aspect of our mental mind. It's not the only one, but it's one. There are obviously non-conscious features of mental life.

Suzan Mazur: Do you see any tie-in between origin of life and cognitive systems engineering/ artificial intelligence?

Vincent Müller: Yes. The artificial cognitive systems approach is exactly to try and understand what actual cognitive systems are like and use that understanding to generate artificial cognitive systems that also have those features. And also to use the understanding of the artificial systems as a test bed for a theory about natural creatures.

Suzan Mazur: Do you see life as non-algorithmic and artificial intelligence as algorithmic?

Vincent Müller: That's a complicated question, again. My view is that anything to do with our digital computer systems is fundamentally algorithmic. If the system that we're trying to generate is not fundamentally algorithmic, then we won't be able to generate it. It seems fairly clear to me that life is not of an algorithmic nature therefore we won't be able to generate it on our computers alone. So artificial life in that sense is not life.

Suzan Mazur: Would you like to make a final point?

Vincent Müller: I have fairly strong views on a couple of the theoretical points we've covered, but I do not claim any expertise on the points for foreseeing the future, particularly the economic parts discussed.

Jaron Lanier

JARON LANIER
(*artist, Peter Sheesley*)

"Hopefully -- and it's hard for man to make that jump -- hopefully at the end [of the film *Insect Gods*] you don't care about the fate of the few. You don't care about the fate of man. You see that civilization has advanced in another way. That it's the roaches that have inherited the Earth. That they have become the gods." -- *Saturday Night Live* comic **Michael O'Donoghue** in conversation with me, 1979

April 3, 2014

Great minds think alike -- that is, late *Saturday Night Live* comedic genius Michael O'Donoghue and Jaron Lanier, the "father of virtual reality" and author of the book, *Who Owns the Future?*.

In the late 70s, O'Donoghue was in pre-production on an end-of-the-world film about the ascendancy of the New York cockroach into a diaphanous creature that would replace humans. The film was an homage to Roger Corman, and O'Donoghue told me it HAD to be shot in black and white and that he had rejected funding for color. Sadly, we lost Michael, and *Insect Gods* never got made.

Two decades later, futurist Jaron Lanier *et al.* wrote to the *New York Times* they had a way to preserve archives for a thousand years that would survive various disaster scenarios, claiming that with a budget of $75K they could implant the robust New York cockroach with a "time capsule" of back copies of the *Times Magazine* and then release a certain volume of the archival insects (eight cubic feet) to breed all over Manhattan. Unlike O'Donoghue's scheme, Lanier's was not intended as black humor.

However, more recent experimenters have demonstrated that commands can be successfully sent from a mobile phone or PC to live cockroaches. So descendants of the Lanier implanted roaches could have been sabotaged, posing an even greater challenge to New York and BEYOND, possibly unleashing O'Donoghue's vision. . .

Lanier, a vulnerable man with a Hitchcock-like profile, Rasta hair and generous eyes has become an international media darling as virtual reality pioneer, musician and visual artist. He is cited as one of the most interesting thinkers alive -- once a goat herder -- with no formal college education, only honorary PhDs. He is also not a physicist, yet is the lead endorsement on physicist Lee Smolin's recent book, *Time Reborn*. (Curiously, the other lead endorsement on the book is not from a physicist either. Brockman book agent common thread here?)

In *Who Owns the Future?* Lanier describes himself as a "humanist softie," and he seems to really enjoy being a father. As we spoke by phone, for instance, Lanier playfully cautioned his young daughter just back from outdoors that the pollen in her hair might sprout. In fact, Lanier has dedicated the book to his daughter and "[t]o everyone my daughter will know as she grows up," saying further, "I hope she will be able to invent her place in a world in which it's normal to find success and fulfillment."

Indeed, Lanier's concerns and pronouncements in *Who Owns the Future?*, now in paperback, are chilling regarding the consolidation of power enabled by the digital world. In our recent conversation he told me that "maximum openness actually turns out to be maximum closedness," referring to the five big tightly-controlled platforms.

We also discussed my interest in "who owns origin of life?" touching on the privately-funded world of protocell development.

I find the most compelling part of Lanier's book his hopeful exploration of the idea of people being financially compensated for their online contribution instead of being "exploited" (the caveat being RAMPED-UP TRACKING):

> "The existence of advanced networking creates the option of directly compensating people for the value they bring to the information space instead of having a giant bureaucracy in the middle, which could only implement an extremely crude and distorting approximation of fairness."

I also adore Lanier's attention to the real nitty gritty, "the end of laundry and never having to wear the same dress twice." Something I once discussed with fashion designer Betsey Johnson. Lanier predicts a countertrend to this technology, as well, *i.e.*, greater reverence for vintage and handmade clothing. Bill Blass would have agreed.

Excerpts of my conversation with Jaron Lanier about the future and who owns it follow.

Suzan Mazur: You've written a book titled, *Who Owns the Future?*. Do you have concerns about a few people, privately funded, acting as intelligent designers of more or less a second genesis on Earth? Also, do you have concerns about what they will create?

Jaron Lanier: I have what I hope is a somewhat nuanced position on that. I know some of the people in the protocell world and think for the most part they're good eggs. That world is somewhat more ethically aware than the computer science world, for instance, which is spying on people, taking advantage of people. Doing damage to the economy.

I am, of course, concerned -- as you are -- with the small number of people making protocells, the extreme control, and that it's being facilitated by a few rich people instead of being publicly funded.

Suzan Mazur: Have you been inside some of the synthetic cell labs?

Jaron Lanier: Sure.

Suzan Mazur: You've been to one or two of them? You've been to Jack Szostak's lab?

Jaron Lanier: I've been to three labs.

Suzan Mazur: Did you have to sign a confidentiality agreement?

Jaron Lanier: I have not signed confidentiality agreements. There might be some that are implied. I'm of the tech and science world.

Suzan Mazur: They trust you.

Jaron Lanier: This notion of very tight control and a very narrow super elite at work is not exclusive to the protocell labs. The whole world is like that right now. The computer cloud that works with artificial intelligence and works with machine learning with giant databases is also restricted to the five big platforms, which are pretty tightly controlled ultimately. The whole world has become one of tiny elites and the rest who are kind of left out. It's a negative trend all around with this being one good example.

It blows my mind that in a very twisted way what seems to people like maximum openness actually turns out to be maximum closedness.

Suzan Mazur: How soon do you think we'll have a protocell?

Jaron Lanier: I don't really know. I'm kind of more interested in trying to figure out some way to develop a better empirical technique so we can really know what we've done. I think there's a tremendous fallacy in the field to think that you can see finality, like in computer programming -- that you think you know what can happen.

Suzan Mazur: Thinking that computer chemistry can replace bench chemistry, for instance?

Jaron Lanier: Using simulations for virtual possibilities is perfectly worthwhile . . . But we can't pretend that the simulation is reliable before really doing the very, very hard work.

We're all familiar with weather prediction. It has gotten better over time. And biology's surely going to be harder than weather to predict. But because it's so much harder to gather data on microbiology, we're free to fantasize that we have more predictive power than we really do. That illusion is the one that really scares me.

Suzan Mazur: Do you see life as algorithmic or non-algorithmic?

Jaron Lanier: I think it's a misleading question because it depends on what you mean by algorithmic. In terms of formal algorithms we study in computer science or mathematics -- these things do not ever really exist in physical reality at all. The very idea of a computer that can be described by an algorithm is a bit of a fantasy. What we do is we create an artificial zone where entropy [information] is suspended and where there's this perfect determinism for a period of time.

It can't last forever. That's what we think of as a computer. And algorithms very quickly correspond to what we can achieve if we do that, but even then it's not perfect. Things will always break down, after a while a cosmic ray will zap them, etc.

333

Suzan Mazur: And so you don't see life as algorithmic.

Jaron Lanier: If by algorithmic you mean the thing that we study in computer science -- it doesn't even exist in reality. . . If by algorithmic you just mean causal, then of course everything is algorithmic. The problem is you're caught between two extreme definitions of algorithmic, and I don't think there's any middle one.

Suzan Mazur: You describe yourself in your book as a "humanist softie" and have also said that we humans may have taken a wrong turn -- implying our turn into the digital world. But that we did it for the right reasons. Other thinkers, Piet Hut, for instance, an astrophysicist at the Institute for Advanced Study, has said that we may have taken a wrong turn to the objective pole with our very focus on science -- which Hut thinks can't be purely objective anyway. Hut says there are other ways of knowing, that science has been only 1% of our human history and that the other 99% -- the subjective pole -- needs further exploration. Would you comment?

Jaron Lanier: I don't think you can really do experiments with consciousness unless you do experiments with aspects of reality. I mean I understand where Piet Hut's coming from certainly. I'm sympathetic with it, but as a practical matter, I don't know what more you do with consciousness other than enjoy it.

Suzan Mazur: David Orban, founder and director of Singularity's Institute for Artificial Intelligence - Europe, told me at a robotics conference in Bergamo a few months ago that people who don't embrace robotics in the future will not be able to survive. Do you agree?

Jaron Lanier: First of all, I think it's the stupidest institute ever. It's purely about this religious fantasy of superiority. The whole basis of it is repulsive. Yet the people there are great friends of mine. I admire them. We have fun together. And I tell them all this to their faces. I've also given talks at Singularity about how ridiculous I think it is.

Here's the problem. They say people won't be able to survive if we don't have robotics. Well, how is that different from saying: Oh, if we don't like the way people are, we'll kill them. What is the difference ultimately?

There's a way in which the new sort of vaguely Asperger-like digital technocrat is absolutely lacking in any self-awareness of ethics or morality. It astounds me, again and again. They're my friends, and we like each other, but I do think it's astonishing.

Suzan Mazur: The European Union is heavily investing in industrial robotics and will retrain workers for jobs lost to robotics. But can it work in the US, which is not a social democracy, where job retraining will be left to the unemployed to shoulder the cost?

Jaron Lanier: I don't even know if it can work in Europe. It can work in the early phase in Europe, perhaps, but you can't have a situation where you pretend that all the people aren't needed for anything and robots do the work. This gets to the illusion of Big Data. The truth is the only way to make machine learning algorithms work is robotics or autonomous systems. And it all depends on what we call Big Data, which means massive contributions from massive numbers of people. Without people creating examples, modifying them and reacting to them and all that, the machines can't work. We're pretending that people are less needed.

Now in the early phase you can train people to work with the robots. If you adhere to the artificial intelligence ideology, then gradually you'll find a way to convince yourself that people aren't needed. But you can't have a 100% welfare state. So even Europe will break eventually. But the US will break first certainly, as we'll find out.

Suzan Mazur: Is job application via computer one of the reasons so many people in the US are unemployed? Do we need to go back to a more human Human Resources Department? Are the robots looking for catch phrases in CVs failing to recognize human ability and capability at the other end of the computer? Do we need to return to a more human system of hiring?

Jaron Lanier: It's absolutely true that mechanization of human resources has been cruel and ridiculous. That's very, very true. A lot of that was driven by liability avoidance where if the machine did it, then a company couldn't be sued for discrimination. We're pretending that we're doing things that are really needed but aren't actually needed. It's a giant illusion. I think the reason there are so few jobs for qualified people is that we're exploiting those people online.

Suzan Mazur: How far along are we in the digital world with the concept of rewarding people for their real contribution?

Jaron Lanier: The problem still is that online there's a way for you to pay for stuff but no way for you to earn your real value. People are adding value to all the cloud algorithms that allow artificial intelligence to operate, but there is as yet no way for those same people to benefit.

Part 6

Rethinking the Circus

"Science must be free to examine what it sees. If you're going to say everyone must follow the Darwinian line, that's not free science." -- **Carl Woese**, in conversation with me, October 2012

Chapter 15

Carl Woese

Evolution's Golden Revolutionary

CARL R. WOESE
(photo, Jason Lindsey, courtesy UI-UC)

"The time has come for biology to enter the nonlinear world."
-- **Carl R. Woese**, "A New Biology for a New Century"

In Memory of Carl Woese, who died December 30, 2012. This is possibly his last feature interview.

In 2012, with the Nobel Prize announcements coming up, I was reminded that there was probably no other scientist but Carl Woese who could write about having "no use for natural selection" and have *Nature* magazine respond with a Nobel endorsement. It was Carl Woese who first identified the archaea and introduced us to horizontal gene transfer. At the time of our October interview, Woese and his University of Illinois, Urbana - Champaign colleagues had also just been awarded an $8M NASA Astrobiology Institute grant to identify the principles of the origin and evolution of life.

Woese told me that his collaboration with physicist Nigel Goldenfeld, a principal investigator on the NAI-funded research, was the most productive one of his science career. In a paper called "Biology's Next Revolution," the two confronted the Darwin issue:

> "Thus, we regard as rather regrettable the conventional concatenation of Darwin's name with evolution, because there are other modalities that must be entertained and which we regard as mandatory during the course of evolutionary time."

I asked Carl Woese how he defined life.

"That's the problem," he said. "We can't."

Woese also said that the fossil records were unreliable.

Freeman Dyson writing in the *New York Review of Books* in 2007 described the quality of Woese's thinking: "Whatever Carl Woese writes, even in a speculative vein, needs to be taken seriously." Dyson termed the Woese hypothesis of a collective network of early life preceding the modern cell a "golden age" and commented that we are now moving forward into another communal stage as a culture.

University of Chicago microbiologist James Shapiro, whose ideas Carl Woese said represented the future, emailed me saying this about Woese:

"Carl Woese was the most important evolution scientist of the 20th Century. He put our picture of living organisms on a solid empirical basis. He discovered a whole new kind of cell. He established the molecular methods for determining phylogenetic relationships. He made it possible to understand the relationships between prokaryotes (bacteria and archaea) and eukaryotes. Any one of these accomplishments would be extraordinary. Altogether they make Carl the most outstanding figure in understanding the diversity of life in well over a century."

Carl Woese described himself as a "molecular biologist turned evolutionist." At the time of our conversation, Woese was Stanley O. Ikenberry Chair and Professor of Microbiology at the University of Illinois, Urbana-Champaign. He taught at UI-UC for almost 50 years.

Dr. Woese's undergraduate degree was in mathematics and physics from Amherst College and his PhD in biophysics from Yale University, where he also did postdoctoral research. For several years in the early 1960s, Woese was a biophysicist at General Electric Research Laboratory.

His honors include the Crafoord Prize in Biosciences (Royal Swedish Academy, 2003); National Medal of Science (2000); Selman A. Waksman Award in Microbiology (National Academy of Sciences, 1997); Leeuwenhoek Medal (Royal Netherlands Academy of Arts and Sciences, 1992); MacArthur Fellow (1984); among others.

He was a member of both the Royal Society and the National Academy of Sciences and author of the book *The Genetic Code: The Molecular Basis for Genetic Expression*, and numerous pivotal scientific papers.

My October 2012 interview with Carl Woese follows.

Suzan Mazur: Congratulations on the recent NASA Astrobiology Institute grant awarded to you, Nigel Goldenfeld and your colleagues to research "the fundamental principles underlying the origin and evolution of life," based on your work over these many decades about life prior to the emergence of modern cells. Why do you think NAI chose to give you and your team $8M, since you are known as a challenger of Darwinian dogma? Is NASA finally acknowledging the Darwin approach is wrong?

Carl Woese: I would hope so because that's very clear from our NASA Astrobiology Institute grant application. I have not seen the reviewers' comments but I've heard that they were quite positive.

It's important to mention others on the team, aside from Nigel Goldenfeld and myself: From UI-UC: Elbert Branscomb, Isaac Cann, Lee DeVille, Bruce Fouke, Rod Mackie, Gary Olsen, Zan Luthey-Schulten, Charles Werth, Rachel Whitaker; from UC-Davis: Scott Dawson; and from Baylor College of Medicine: Philip Hastings and Susan Rosenberg.

Geologist Bruce Fouke's work is going to be a large part of the outreach component of this NAI grant. Bruce works at Yellowstone National Park and has a wonderful program going with those interested in the scientific side of Yellowstone. He takes small groups of people up there to observe geological formations, etc. They learn a lot from this.

Suzan Mazur: Patterns of organization.

Carl Woese: NASA is big on outreach, as you know. Both Bruce Fouke and Isaac Cann are very concerned with teaching the next generation of scientists. Isaac is an archacaologist. Archacaologist in the sense of archaea, one of the three domains of life.

Suzan Mazur: I was also excited to see that you're going to put together a free online course for students. Hopefully others will take that course as well, like media editors, etc.

Carl Woese: Yes, I was talking just yesterday with Isaac Cann, for example, one of the principal investigators of the project who's very keen on this sort of thing. There is an online course being put together by members of the astrobiology institute here and at other facilities on this campus. It's a genuine teaching effort.

Suzan Mazur: So with this grant you're going to be figuring out the general principles of life. How do you define life?

Carl Woese: That's the problem. We can't. We have yet to answer central questions about the origin of life. We have yet to get more direct evidence for what I call a pre-Darwinian condition, a progenote condition of life. That's one of the things we're working on, trying to get as much direct evidence as we can. Obviously, since this is a stage in evolutionary space of three or more billion years ago, we're not going to get much in the way of direct evidence. We can get fossil record, but that's not reliable. You have to infer everything you can from intelligent insightful analysis of genome sequence data.

Suzan Mazur: What do you consider evolution?

Carl Woese: Evolution is actually what biology should be. What is biology? Is it some under-the-microscope description of forms? It can't be that. Evolution is a process. It is the process which we now call biology which is very static. Evolution, however, is dynamic. And we have to understand what rules that dynamic follows.

Suzan Mazur: There's been recent controversy about whether astrobiology is a real science or a way for scientists to secure funding. Antonio Lazcano, former president of ISSOL (the International Society for the Study of the Origin of Life and Astrobiology Society) wrote in August in *Nature* that "Depending on who you speak to, astrobiology seems to include everything from the chemical composition of the interstellar medium to the origin and evolution of intelligence, society and technology -- as if the Universe is following an inevitable upward linear path leading from the Big Bang to the appearance of life and civilizations capable of communication." Would you comment on that?

Carl Woese: Yes. It's not a linear path. . . And you don't use the word upward. The universe is a process.

Alfred North Whitehead said with biology and other things, we are not dealing with a procession of forms . . . we are dealing with the form or forms of process. In that distinction lies the essence of what this astrobiology institute here in Illinois is all about. We are going to really study the evolutionary origin of life. Anybody who looks at biology now, with a one percent exception, can feel -- anyone with biological intuition can feel -- that life is an evolutionary process. As I said, it's not just a procession of forms.

Suzan Mazur: Antonio Lazcano has also been quoted as saying regarding the investigation into early life, that we cannot jump the molecular hurdle prior to the appearance of proteins. What are some of the "principles of life" you and your colleagues are looking to confirm?

Carl Woese: We're certainly not looking for proteins to be there in the absence of nucleic acid, I'll tell you that. If I were to tell you what principles we were looking for, there would no longer be a question. We have to try to discover the dynamic of the process of evolution, of the evolutionary process. For this the biologist needs a lot of help, particularly from mathematicians and physicists who are used to dealing with complex systems. Systems so complex they iterate by themselves.

Suzan Mazur: How much are you relying on computer simulation for your work?

Carl Woese: I'm relying on my collaborators for the computer simulation work, but I'm making inclusive judgment about what they do. So there are good, productive discussions among us.

Suzan Mazur: How much of NASA's interest in investigating origin of life would you say is based on creating synthetic life for commercial applications and how much knowledge-driven?

Carl Woese: There are many people who like it because they can patent things and make a lot of money.

Suzan Mazur: But do you think there is a genuine interest at NASA in getting to the bottom of the origin of life?

Carl Woese: There's always been. In the 1970s, NASA and NIH jointly sponsored my research for The Third Domain.

Suzan Mazur: Great. Not as upsetting to the Darwinists as the current investigation.

Carl Woese: No. There was something new in my work, and it was at a meeting in Paris that Dick Young, the first head of NASA's "Planetary Biology" section, stepped forward and offered help.

Suzan Mazur: Was it a large grant?

Carl Woese: By today's standards, it was minuscule.

Suzan Mazur: Do you have any concerns about the creation of a protocell?

Carl Woese: Oh yes. There are some, as you know, Craig Venter is beating the drum on this all the time, just to be at the forefront. Power. . . .

I have concerns about scientists thinking that they're God when it comes to biology. Scientists should be trying to study the experiments that nature has already done in the form of the evolutionary process.

Suzan Mazur: You've described the "disconnect between Darwinists, who had taken over evolution, and microbiologists, who had no use for Darwinian natural selection." Do you have anything to say about the recent decision of *Huffington Post* to block publication of microbiologist James Shapiro's response to Darwinist Jerry Coyne following Coyne's attack on Shapiro's thinking about a reduced role for natural selection in evolution?

Carl Woese: I think that's immoral. Science must be free to examine what it sees. If you're going to say everyone must follow the Darwinian line, that's not free science. *Huffington Post* has gone from right to left to right to left. I don't know where it is now. This doesn't belong in science.

I think Shapiro has got his finger on the future. He sees that we should be studying regulation. Epigenetics is very important.

Suzan Mazur: You've also noted that Darwin's thinking on common descent is "chiefly grounded on analogy" and that the evolution now emerging is not coming from Darwin, commenting further that:

> "A future biology cannot be built within the conceptual superstructures of the past. The old superstructure has to be replaced by a new one before the holistic problems of biology can emerge as biology's new mainstream."

Do you expect Darwin to go the way of Freud as "biology enters the nonlinear world" and evolution is redefined?

Carl Woese: It could well do that. I've maintained for a long time up until the end of the 20th century that the problem of the evolutionary process is a problem before its time. Darwin was trying to get personal credit by barging in. Conceptual thought about evolution was laid down first by people like Buffon and Darwin's own grandfather, Erasmus Darwin -- whom Darwin never mentions in *The Origin of Species*, except in a footnote when he was forced in the third edition to add it to the footer of the preface.

He named him in a dismissive way. He basically said, oh yes, a lot of people thought of that and named people like Buffon and Lamarck. But he didn't name his own grandfather, Erasmus Darwin, except to say his grandfather had the same wrong ideas as Lamarck and Goethe. And he didn't say what they were or what his objection to them was. He wanted to distance himself from his grandfather as much as he could. . . ."

Suzan Mazur: Nigel Goldenfeld has been lecturing about "three dynamical regimes." Is he referring to what you outlined in your 2006 paper -- weak communal evolution, strong communal evolution and individual evolution?

Carl Woese: Yes, I believe he is. I'm almost positive it is. And in something I wrote on archaea, I speak of the evolution of individuality. There was a communal stage to begin with. This is what I usually call the progenote.

Three domains is what I use. I wrote something in 2004 for *Microbiology and Molecular Biology Review*. Freeman Dyson was taken with that and asked permission to use it in something he wrote for *New York Review of Books*.

Nigel's trying to define for physicists what these three domains are. This is one of the few points in which I differ from Nigel Goldenfeld. It's a friendly kind of discussional difference.

Suzan Mazur: It's wonderful that you're making these breakthroughs without encountering too much hostility from the classical biology community.

Carl Woese: But I have not overthrown the hegemony of the culture of Darwin.

Suzan Mazur: Do you have any closing thoughts?

Carl Woese: Yes, I do not like people saying that atheism is based on science, because it's not. It's an alien invasion of science.

We Need a Theory of Life

NIGEL GOLDENFELD
(*photo by Lou Mcclellan, courtesy N. Goldenfeld*)

June 9, 2014

I was intrigued when Carl Woese told me his collaboration with University of Illinois physicist Nigel Goldenfeld was the most productive one of his entire career, and was pleased to finally run into Goldenfeld last September at lunch in the courtyard of the Santa Fe Institute. But there was little time to talk since I had a meeting in town and he was about to give an "Emergence" lecture.

Goldenfeld, British by birth, speaks faster and more distinctly than most humans, and moves at equally fascinating speed. I stayed long enough at Goldenfeld's lively talk to see him demonstrate various ways carbon organizes, watching first as he pushed over a wooden chair and then got a standing-room-only gathering of some of the world's edgiest scientists to happily agree to put their fingers in the air in an attempt to hold up the ceiling.

More recently while traveling in the Midwest, I planned a visit to the NASA Institute for Universal Biology at the University of Illinois, Urbana-Champaign, where Goldenfeld is director. But flights through Chicago were canceled that day due to a fire at O'Hare, and I was unable to meet with Goldenfeld and the IUB team: Elbert Branscomb, Bruce Fouke, Zan Luthey-Schulten, Charles Werth *et al.*, who are now two years into a five-year $8M project figuring out the principles of the origin and evolution of life.

Goldenfeld was returning from Washington, DC the day of the airport fire and rerouted through Indiana, taking a limo the rest of the way home. He was gracious about our missed appointment and said he and the IUB team clearly don't have all the answers yet, adding that what makes the task particularly tricky is the fact that most life is microbial and that "humans are not representative of general principles!"

As a condensed matter physicist, Goldenfeld's main interest is pattern formation and the processes involved in pattern formation. He researches everything from snowflakes to geological formations to the stockmarket. He and his team have captured gorgeous images of some of these formations in progress at Yellowstone National Park.

A separate curiosity of Goldenfeld's is why so many "offspring" of scientists have autism.

Einstein, Newton, Tesla, and other science giants are now thought to have had Asperger syndrome, a form of autism. And Darwin, as Scott Stossel points out in his recent book on anxiety, exhibited nine of the thirteen indicators of panic disorder (Stossel says only four are needed to qualify).

Other research interests include high temperature superconductors, turbulence, critical phenomena, polymers and liquid crystals.

Goldenfeld's BA, natural sciences, and PhD, theoretical physics, are both from Cambridge University. He studied there with condensed matter physicist Sam Edwards, who was knighted for his service to science. Goldenfeld says Edwards deserved a Nobel as well -- as did Carl Woese.

Goldenfeld did his postdoc at the Institute for Theoretical Physics in Santa Barbara. He has been a visiting scientist at Stanford University and has taught at Cambridge.

Aside from his role at the Institute for Universal Biology, Nigel Goldenfeld is director of the Biocomplexity Group at UI-UC's Institute for Genomic Biology. He has been funded by the National Science Foundation for the last 30 years with additional grants from the Department of Energy. He rejects "overtly military" agency funding.

Goldenfeld is the recipient of the Xerox Award, A. Nordsieck Award, the National Science Foundation Creativity Award, among others.

He is a Fellow of the American Physical Society and the American Academy of Arts and Sciences, was an Alfred P. Sloan Foundation Fellow (1987 - 1991), and is a member of the National Academy of Sciences.

Some of the journal editorial boards he serves on include: *The Philosophical Transactions of the Royal Society*, *Physical Biology*, and the *International Journal of Theoretical and Applied Finance*.

He is the author of: *Lectures on Phase Transitions and the Renormalization Group*, and co-author with Paul Goldbart and David Sherrington of *Stealing the Gold: A Celebration of the Pioneering Physics of Sam Edwards*.

My phone interview with Nigel Goldenfeld follows.

Suzan Mazur: Carl Woese told me that his collaboration with you was the most productive of his scientific career. What was special about the collaboration for you?

Nigel Goldenfeld: Carl was very generous in saying that. It certainly was a very special collaboration for me. It isn't often that one gets to work with someone with Carl's breadth and intellect.

Carl was, like me, a physicist by training. [Carl Woese, PhD Yale, 1953, biophysics] He wasn't a theoretical physicist but he had a strong intuition, so we saw things in a similar way. We very quickly found common ground. There didn't need to be a lot of explanation between us. That was quite special because one of the potential obstacles to interdisciplinary collaboration is a language barrier between say biologists and physicists.

Others complain about this but I've actually not run into that too much. People I've worked with in the biological sciences have all been very available to me, and prepared to explain what they're doing in simple language that a physicist can understand. Carl was really open to concepts that I think were important to me as a fairly naïve scientist coming in. He was already thinking about the questions that I found natural to ask. They were quite conceptual questions, but concrete ones, and could be answered scientifically. It was clear Carl had been thinking about these questions for decades.

Carl's papers from the 1970s are still fresh because many of the ideas in them have essentially not been picked up by other people. These ideas may have been overlooked because they were not comprehensible to people at the time. They seem very relevant now.

Suzan Mazur: Carl Woese in that same interview with me about two months before he died lamented that he had not managed to overthrow "the hegemony of the culture of Darwin." His view was that a "future biology" couldn't be built within the past structure, that it had to be replaced. And you've both noted that it is "rather regrettable" that Darwin's name is so synonymous with evolution because "there are other modalities that must be entertained."

What are some of these other modalities?

Nigel Goldenfeld: The prejudice that one runs into often (but not always) when one talks about evolution is that evolution really is dominated by what's called vertical gene transfer, where first of all evolution is essentially all about genes -- operations and effects -- evolution occurs because mutations in genes are transmitted to offspring and there's a standard natural selection approach.

There isn't any doubt about this role of population dynamics in biology, by which I mean that entities can replicate and perhaps dominate a population. But the ways in which genetic novelty can emerge and be transmitted are much broader than that, and provide additional channels through which the well-understood mechanisms of population biology can act.

For example, Lamarckian evolution is definitely something that happens in the biological world. As a matter of fact, Darwin himself did not exclude that possibility in *Origins of Species*. But most people, most lay people and most biologists would regard the idea of something Lamarckian happening as something not within the standard picture of evolution. Lamarckian in this sense really means the inheritance of acquired characteristics.

Suzan Mazur: Epigenetics.

Nigel Goldenfeld: Yes. But there is an even simpler, less controversial example -- horizontal gene transfer.

If you are a bacterium and you receive a plasmid and some genes and you can then later express those genes and transmit them to the daughter cells, that's an absolutely manifest example of something that was acquired in the lifetime of an organism, the result of its environmental interactions that gets transmitted to its offspring. We know that horizontal gene transfer is a significant force in microbial evolution.

Suzan Mazur: So it's happening inside of us, and outside of us as well, horizontal gene transfer because of our microbiota.

351

Nigel Goldenfeld: Yes. Exactly. That's exactly right. It's responsible in part for the proliferation of antibiotic resistance genes. Antibiotics are becoming less effective as a result of horizontal gene transfer.

Suzan Mazur: So two of the other modalities are horizontal gene transfer and Lamarckism, epigenetics.

Nigel Goldenfeld: Exactly. There's epigenetics and there are other questions. For example, how do we really know that the standard dogma is true that there's genetic variation which occurs as a purely random process and whatever is present in that variation is then selected by the environment? That would be the standard picture that the evolutionary dynamics itself is random and possible for selection.

Perhaps the best evidence for that is actually a justly-famous experiment: the Luria-Delbruck experiment. It's a brilliant experiment but because of the way the experiment is done, it does not close the book on the idea that organisms and cells can sense environmental stress, and indeed may be able to respond in ways that are not random.

I should add a clarification about what is meant by random, because it frequently gets confused in this sort of discussion. It is generally accepted that mutation rates can be up-regulated due to stress. It is also generally accepted that mutation rates can be site-dependent. Neither of these phenomenon relate to randomness in this context: the issue is whether the rate is affected by the environment. As a physicist, for me an experiment that gets a zero or non-zero result is not adequate. If the result is zero (which would be the standard assumption), then one would actually want to bound the result systematically. If not, one would like to measure the dependence as a function of environmental stress. I don't think these sorts of tests have yet been done, and that is why I think there is room to explore this question. Precise tests are beginning to be possible with physics-based tools that can probe living cells in real time.

In short, nobody is saying (well, I'm not at least) that evolution is

352

a teleological process leaning toward a particular end, but there's certainly logically a possibility that the evolutionary process is responsive to the environment in a way that is not random. That would not be part of standard Darwinian evolution and population genetics.

Suzan Mazur: Do you differ from Carl Woese on replacing the Modern Synthesis? Woese wrote the following in "A New Biology for a New Century":

> "A future biology cannot be built within the conceptual superstructure of the past. The old superstructure has to be replaced by a new one before the holistic problems of biology can emerge as biology's new mainstream."

Nigel Goldenfeld: Our collaborative position was that the Modern Synthesis is simply not enough, population genetics is not a full account of the evolution process because it manifestly does not describe evolution before genes, it does not describe evolution before there were species and the lineages. The Modern Synthesis wasn't designed to do so. Amusingly, Huxley, one of the founders of the Modern Synthesis explicitly wrote that it would not apply to bacteria, but this statement was written before it became clear that the mechanisms of heredity operate in bacteria as in other "higher" organisms.

Usually the way science proceeds is not by demolishing and putting in its place something that's not organically related to it. There's usually an expansion of the theory and a retaining of what still applies, in just the same way that Einstein's theory of relativity supplants Newtonian gravity. It doesn't mean that Newtonian gravity no longer is relevant in the situations for which it is applicable. But Einstein's theory has a greater generality and can apply to physical situations where Newtonian gravity breaks down. The same thing is true with quantum mechanics and classical mechanics.

So our perspective is that if one is looking at origin of life issues, then population genetics, which is really designed for looking at modern life, as it were, is not a full account of what happens generally in evolution and specifically at the early stages of life where genes and species and individual organisms, etc., had yet to emerge. But I don't think there's a real contradiction there, and I don't think this idea is particularly surprising or controversial.

Let me be more precise about the way in which new theory emerges. You might say we have to replace Newtonian gravity by a more general theory, and it has been replaced by Einstein's general theory of relativity. But Newtonian gravity is a perfectly good example of the limit of general relativity in certain circumstances when the gravitational fields are weak and so forth. So I think that's what Carl meant and that's certainly what we were working to try and understand. The Modern Synthesis doesn't address and doesn't claim to address issues of how do living systems even arise in the first place and how do you account for the very existence of life as a phenomenon.

The Modern Synthesis says if you have genes and if you have selection, etc., it gives you a mathematical account of the frequencies of alleles in the populations, and in many situations one could do a calculation. To the extent that can be tested, it does a good job and that's what it was intended to do. That's not the same thing as a comprehensive theory of life as a phenomenon, of the emergence of life, the time before the distinction between genotype and phenotype arose.

So it's replacement with a deeper level of understanding, but it still allows for the old theory in the situations where the old theory applies.

The same thing is true of quantum mechanics. Quantum mechanics was a complete departure from the world of classical physics, yet when we look at systems on a large enough scale and with appropriate energies, the predictions of quantum mechanics are identical to the predictions of classical mechanics, and we can derive classical mechanics out of some limit of quantum mechanics.

That more general theory has a different perspective and a different framework. It has that extra level of generality that we can apply to other situations. I think the same will be true with our understanding of living systems.

Suzan Mazur: Carl Woese also told me that the Last Universal Common Ancestor was not anything material and that evolution is not a "procession of forms," it's more a procession of processes and that physicists and mathematicians are important to the understanding of this new biology. Can you please clarify this idea that LUCA was not something material?

Nigel Goldenfeld: What Carl meant was this. The nature of the evolutionary process we think was different than it is today. Early life was much more collective, much more communal than it is today, particularly the core cellular machinery such as translational machinery, etc., which was horizontally transferred.

We don't know how that happened. It may well have been that there was massive endosymbiosis, meaning organisms were very porous and could crash into each other and absorb each other on a massive scale and that's how cellular functions were transmitted.

That was an idea Carl proposed in the 1970s in one of his papers. But when you have a system which operates in that way, the dynamics -- how the thing changes in time -- obeys very different mathematical principles than what happens post-LUCA when you have vertically-dominated evolution.

An example of this is, suppose you ask what is the defining characteristic of evolving systems, whether they're biologically evolving with genes, etc., or they are other systems you might want to consider living in some sense (for example, financial markets perhaps). The perspective that Carl and I put forward in our last paper together is that life is inherently self-referential. In other words, it's like a computer program which can constantly overwrite the program itself as it runs. The notion that program is the data and the data is the program, the idea that life has a dynamics where the rules that govern life are themselves changed by the rules -- that's self-referentiality. It gives you a kind of description of the physical system that behaves in a mathematical way but is quite unlike any other system that we've ever studied.

It's not like the laws of classical mechanics or the laws of quantum mechanics or the laws of statistical mechanics. Trying to understand the evolutionary process in a more general way requires us to think in a different way mathematically than the way we've thought in the past. I do work with and talk to mathematicians about these sorts of issues.

Suzan Mazur: You're particularly interested in patterns of organization. From snowflakes to the stockmarket to research on autism. Can you tell me a little bit about your ideas regarding each of these areas of research and how they are linked?

Nigel Goldenfeld: I don't know that they are all linked. They're all linked because I'm interested in all of them. It's not that I think there's an overarching principle that connects to say patterns in nature to patterns in ecology to patterns in the evolution of life to autism. That being said, there are some commonalities.

One thing is that systems that are out of equilibrium are much more interesting than systems that are in equilibrium. Systems in equilibrium lapse into perfect states like crystals, etc. Systems that are out of equilibrium are messy and produce turbulence and swirly clouds and human beings and galaxies and strange patterns in space and time. It's interesting to try and understand this, it's my main intellectual interest.

Suzan Mazur: How does autism fit in?

Nigel Goldenfeld: It doesn't really fit in at all. I was interested in autism for many reasons. One is that the autism spectrum conditions are genetic and are more common in offspring of physicists and other scientists and mathematicians than the general population. It's an interesting and important phenomenon.

Suzan Mazur: Asperger syndrome.

Nigel Goldenfeld: Right. When I was on sabbatical at Cambridge years ago I ran into Simon Baron-Cohen, director of the autism research center, and I was able to contribute mainly in a technical way to the research they were doing by suggesting new ways to analyze their data. It's not that there's a grand plan there. But it's sort of connected to my other interests.

Suzan Mazur: Antonio Lima-de-Faria in his 1988 book *Evolution without Selection* gives a lot of space to crystal organization and growth and notes that the patterns displayed by plants and animals are already present in minerals. What are your thoughts about that in light of your interest in the relationship between geochemistry and metabolism at the scale of atoms in the origin and evolution of life?

Nigel Goldenfeld: I don't remember the parts of that book you're talking about in detail, but it's certainly true that there are very common patterns and motifs that you see over and over again in physical systems and living systems. It's not surprising, because in both those cases what's happening is that you have some variable that may be diffusing, say like heat in a growing crystal or maybe some chemical in the case of something biological or maybe cells diffusing in an environment of food. Those processes have a mathematical aspect but there's not a deeper principle at work.

Suzan Mazur: Lima-de-Faria also refers to bacteria having magnetite particles in their cells and can orient in a magnetic field because of this. Can you say anything more?

Nigel Goldenfeld: I've heard of this but I don't know anything about it. Ralph Wolfe, a wonderful microbiologist who worked with Carl Woese in the discovery of the archaea, I know he's interested in this phenomenon. It's not true of all bacteria as far as I know.

Suzan Mazur: Do we need a new language for the world around us? Noam Chomsky has said society encourages us to identify what we're looking at as trees, rivers, dogs, etc., when what we're really seeing is something else. So do we need a new language?

Nigel Goldenfeld: Sometimes there are things that are hard to understand because we don't have good concepts to present them and so we have to struggle to find the right mathematical structures to represent them. I don't think we need a new language or know what form that would take. I think the scientific methods will either produce the right language or generate the situations where the need is so acute that it will appear automatically.

Suzan Mazur: Do you see a difference between life and non-life?

Nigel Goldenfeld: Yes. . . . I think there's a very big difference. The difference is this. When we talk about life, we talk about an abstraction that's occurred at an emergent level of organization. When we look at the individual atoms and molecules that make up life, there's no difference between living and non-living things. On the emergent level, they have different behavior, and the reason they have different behavior is because they have this ability to have a self-referential dynamics. They behave in ways that are very difficult to predict for that reason. So living systems represent a different class of dynamical systems. But at the smallest possible levels, there is absolutely no difference.

Suzan Mazur: Are the factors too overwhelmingly numerous to really come up with a reasonable approach to origin of life? Is it all too relational-based, too unpredictable, to actually build a self-reproducing protocell? Is origin of life beyond science?

Nigel Goldenfeld: No. I think there are two questions there. One, why does life as a phenomenon even exist? What I mean by that is the following. If you think of anything else that we think we understand -- my pet example is superconductivity because I'm a condensed matter physicist -- there are two levels of understanding.

The first one is why the phenomenon exists and is it based on very general principles of symmetry and conservation, conservation of energy, gauge invariance, etc. For superconductivity, that level of understanding came very late, and its absence prevented us from discovering many of the core features of superconductivity for several decades. I'm thinking of the expulsion of magnetic fields, for example, or the existence of what is known as type-II superconductivity and the existence of magnetic vortex lines. **Once we understand why the thing can exist in the first place,**

then we can understand how it is instantiated in any particular system.

For example, for classic superconductors, we understand the connections between the electrons and the sound waves in that particular material and how it is all put together quantum mechanically with electricity and magnetism. So for these materials we understand superconductivity both at a general level from the point of view of symmetry and symmetry breaking, and we also understand it at a level of specific behavior of the atoms and molecules in the crystal structure. For the high temperature superconductors, we still lack this level of understanding of instantiation, although we know they follow the same basic principles as the classic superconductors.

When we talk about living systems we don't have the general understanding, the symmetry-based understanding. We're unable to say, yes, this is a phenomenon and should exist. We need to develop a fundamental understanding of living systems, and from that we can go on to ask, well, given that life exists, how does one instantiate it? How does one make examples in the lab as it were? Up to now, most biological research has been akin to studying superconductors from the materials perspective only, lacking the more general symmetry-based level of understanding. I believe that just as the absence of a general theory held back the development of superconductivity as a quantitative science for decades, the absence of a general theory of life is holding back our understanding of biology. Where this manifests itself mainly is in how we intervene with biological systems, medically and ecologically. The emergence of drug resistance in cancer, or herbicide resistance for that matter, are both examples of our chronic inability to control these systems.

You asked if the origin of life is beyond science. Jack Szostak and others are trying to make synthetic cells, etc. I think science can constrain the environment where life on Earth first arose and in principle be predictive about what sort of life might exist elsewhere in the universe and what might be the properties and biosyntheses. It's not beyond science. It's a question that science can definitely answer.

Pier Luigi Luisi

Origin of Life "Mindstorms" Needed

PIER LUIGI LUISI
(photo, courtesy P.L. Luisi)

"Of dreams we live." -- **Pier Luigi Luisi**

The work of Pier Luigi Luisi continues in the best tradition of Italian visionaries of science and the humanities, Alberti, da Vinci, Galileo, Michelangelo. Luisi is a biochemist, long associated with ETH-Zurich (Swiss Federal Institute of Technology) and with the University of Rome3 -- where he teaches and has served as chair of the university's biochemistry department.

But his Luisi lab at Rome3 has now been forced to wind down, due to lack of funding, and its current research on origin of life, cell models and the self-organization and self-reproduction of chemical and biological systems takes place without his hands-on direction as Luisi increasingly devotes his time to writing books, lecturing and traveling.

I had an opportunity to meet Pier Luigi Luisi -- friends call him "Luigi" -- at the 2013 Scuola Autopoietica del Mediterraneo conference in Sardinia where he and theoretical biologist Stu Kauffman were the two guest speakers.

Pier Luigi Luisi, a tall, graceful man, is revered in science and other cultural circles. Luisi has been particularly masterful at organizing science conferences with a cross-cultural approach in Europe -- 20 or so of them in the last 10 years.

For example, Luisi is famous for his annual Cortona Week. He describes Cortona Week as "a living organism" integrating science and the humanities with a "gallery of original, often fantastic and occasionally odd, strange speakers." His aim has been to showcase "internally richer" future world-class leaders.

Pier Luigi Luisi has published 11 books, among them: *The Systems View of Life*, co-author Fritjof Capra; *The Emergence of Life*: *From Chemical Origins to Synthetic Biology*; *Mind and Life*; *Giant Vesicles* (P.L. Luisi, P. Walde (eds.)); *Chemical Synthetic Biology* (P.L. Luisi, C. Chiarabelli (eds.)); *The Minimal Cell* (P.L. Luisi, P. Stano (eds.)); and several Italian literature books -- including for children and young adults. Luisi is the author of 500 scientific papers.

His PhD is in chemistry, *summa cum laude*, from the University of Pisa. Luisi did postdoctoral work at the University of St. Petersburg in the former Soviet Union and continued his research at the University of Uppsala, the Macromolecular Center Strasbourg, and was Research Fellow at the Institute of Molecular Biology in Eugene, Oregon. He was *Extraordinarius* of Macromolecular Chemistry at ETH-Zurich (Swiss Federal Institute of Technology), founded the Swiss Colloid Group, and co-founded (with P. Pino and J. Meissner) the Institute of Polymers at ETH-Zurich, where he served three times as chair.

My interview with Pier Luigi Luisi follows, updated from our

December 2012 conversation.

Suzan Mazur: You co-organized the San Sebastian origin of life conference in 2009. What are some of the significant developments since then? Would you say the origin of life field is now "exploding"?

Pier Luigi Luisi: No, it is not exploding, except for some nice technical advances. Conceptually it is rather in a stall situation. The main point about origin of life is that we cannot yet conceive how that happened.

We all accept the narrative according to which life arose from inanimate matter, and we have several competitive hypotheses to explain how this really happened, but each of them contains holes and unproven assumptions. And in terms of filling the holes, conceptually or experimentally, in terms of "seeing" the light, I haven't seen in the last few years any really satisfactory breakthroughs.

Suzan Mazur: Is the problem not enough funding to do the research?

Pier Luigi Luisi: No, no, no, it is not. Although funding is always a consideration, the real problem is, I would say, in our mind and intelligence, in the lack of capability to conceive in real chemistry how the passage from nonlife to life arose. The fact that we have several perspectives in competition means that none of them is really convincing.

Not at all convincing to me is, for example, the RNA world, a perspective I find full of conceptual flaws, at least in the doc version. This is the theory by which self-replicating RNA (sr-RNA) arose by itself, began to self-reproduce more or less indefinitely until it evolved -- one wonders why -- into ribozymes, capable then to catalyze the synthesis of nucleic acid and proteins.

Nobody in the field has answered the preliminary question, where the sr-RNA came from. And, even if we find one such sr-RNA, there are lots of chemistry constraints: for example,

you cannot make self-replication with one single molecule. You need at least a binary complex to start replication, with a concentration to overwhelm the dissociation forces.

Simple calculations show that you need at least 10^{-12} M of the replicator, our RNA, to start with. This is a lot, in terms of the number of molecules in one liter, or in one microliter. So, you need a metabolism to make RNA, even before you start scheming with the RNA world.

And even granted that, the story of sr-RNA evolving into ribozymes is also a chemical non-sense. Let us suppose, to start with 1 µM solution of the sr-RNA having 50,000 Dalton, replicating nonlinearly (2, 4, 8, 16. . .) in a small 100 l warm pond for, say, 20 generations. You would end up with 1 M solution, and for that you would need 50 tons of nucleotides -- clearly unrealistic. And finally the ribozyme would possibly couple the peptide bond, but so what?

The real problem is to make ordered sequences of amino acids, and of course ordered sequences of nucleic acids -- and on that the prebiotic RNA world is absolutely silent. But this view of the prebiotic RNA world is still the most popular. I think it is a case of social science psychology more than science itself.

Suzan Mazur: It's remarkable to hear you say that.

Pier Luigi Luisi: Yes I know, and sometimes people think I'm a little too negative and pessimistic, but it is not so.

I simply think we need a new start, a kind of beginner's mind revolution. In particular, we have to come away from the DNA or RNA-centric view of life. This is one of the main reasons why I organized the meeting on Open Questions on the Origin of Life (OQOL) in San Sebastian and again in Nara, Japan in July 2014 -- to call attention to alternative ways of asking questions.

Suzan Mazur: Are you currently coordinating a European COST action for another origin of life conference?

Pier Luigi Luisi: No, I limit myself to the OQOL workshops. As I said, I think it is useful to focus on what we do not yet understand and ask why we don't. And also change the kind of questions we ask.

Suzan Mazur: You're not comfortable with computer chemistry.

Pier Luigi Luisi: We wrote several of our papers on minimal cells with the help of theoreticians and computer scientists. The assistance of these theoreticians and computer model experts can be very important -- but only when they work hand-in-hand with experimentalists.

Suzan Mazur: Would you describe your current research?

Pier Luigi Luisi: It is fair to say that I am now retired. Some years ago I retired from ETH-Zurich, after 33 years of work there. I have now retired from the University of Rome3, where I worked 10 years. I still teach, but I do not have a lab anymore -- a big loss indeed. And here in Italy there are no millionaires to openly support such research.

I can tell you what I was doing previously. Two main areas of research linked to each other: origin of life and synthetic biology (SB). The latter is what we call "chemical synthetic biology," a kind of synthetic biology that does not use genetic manipulation. One research project was the construction of what we call minimal cells.

A minimal cell is a cell which has minimal and sufficient complexity, *i.e.*, the minimal and sufficient number of components to be alive. Synthetic biology is connected to the origin of life because with origin of life we are trying to understand the structure of the initial, simplest protocells.

Consider that an average microbial cell on Earth contains several thousand genes and a few other thousand components -- an enormous complexity. It is precisely this complexity that elicits the question: Is this complexity necessary for life or can we make life with a much simpler structure?

Suzan Mazur: Building from bottom up.

Pier Luisi Luisi: Yes, but not completely from bottom up because extant nucleic acid and proteins are used. In that sense it is not really origin of life because the starting point is half way. The components go into a shell, a liposome. The investigation is figuring out what the minimum and sufficient number of enzymes and nucleic acids are that can transform the liposomes in cell models to display some of the basic properties of a biological cell.

Suzan Mazur: But you consider this a protocell.

Pier Luigi Luisi: Yes, we can call it a protocell. It's a very vague term and indicates a vast family of things. It's something prior to the first fully-fledged biological cell, which is capable of three things -- self-maintenance, self-reproduction and evolution.

In our last achievement, before closing the shop, we were able to prepare liposomes containing ribosomes and capable of expressing proteins -- a very basic property of biological cells. Also, we found conditions by which biopolymers spontaneously over-concentrate inside liposomes (what we call over-crowding) thus solving, I believe, the old problem of how to reach the concentration threshold inside compartments. But the "living cell" is still far away. By that I mean a minimal cell also capable of self-maintenance (homeostasis) and/or self-reproduction.

I had to stop work last year on this cell, but the research goes on in other labs around the world. I am thinking of Yomo and Ueda in Japan and some people in the US.

Suzan Mazur: You mention Yomo in Japan, is that Tetsuya Yomo at Osaka University?

Pier Luigi Luisi: Yes. Tetsuya is a good friend of mine. Very smart. Very intelligent person.

Suzan Mazur: So the replication is something that is down the road.

Pier Luigi Luisi: It is down the road.

Professor Ueda is the one in Japan who discovered this minimal enzyme kit, which he called the pure system. He's trying to modify this minimal set of enzymes so that the set of enzymes reproduces itself. The realization of this, judging from his very latest work, is still out of sight.

Suzan Mazur: How many labs are working on the protocell?

Pier Luigi Luisi: I know of five, six, maybe seven -- the most in Japan. And, of course, Jack Szostak in Boston, but with his emphasis on the RNA world. And I hope that Dave Deamer [now working with NASA's Winona Vercoutere through a grant from US businessman Harry Lonsdale] is still thinking about the minimal cell.

So this is one direction.

The second direction is in chemical synthetic biology, an area we call "never born proteins." An important question in the origin of life in general, in evolution, is why did nature do things in one way and not the other? Why this and not that? One of these questions pertains to proteins.

Now the proteins of our life are only a few thousand billion. You would say this is an enormous number. Yes, but this is a ridiculously small number when compared to the possible number of proteins. I used to say that the known number of proteins corresponds roughly to a grain of sand in the Sahara, compared to all the sand of the Sahara.

How did these few proteins come about? These, and not some others?

In our lab we constructed proteins which were "not born" -- did not exist previously. We were comparing them to the proteins that do exist on Earth -- this, in order to understand whether our proteins on Earth had something basically different from the random library of never born proteins.

It's easy to make a library of proteins at the level of their DNA or RNA, but it's very difficult to obtain these proteins in a milligram or microgram amount in pure form. In this sense our research has been a half failure. The few that we made were proteins completely similar to the proteins of our Earth -- folded, thermodynamically stable and soluble in water. Proteins that can be reversibly denatured.

Let me add a couple of lines about my research on the origin of life. In the last years of my activity in the lab we concentrated on simple dipeptides displaying catalytic activity. **You see, more generally, one can say that the prerequisite for the origin of life is the onset of kinetic control -- namely catalysis -- something which permits a departure from thermodynamic control.**

There are peptides, even prebiotic ones, which are catalytically active as proteases. We were working with Ser-His, a dipeptide that can catalyze the synthesis of the peptide bond, as well as the synthesis of oligonucleotides.

You can use these simple compounds, in principle, to synthesize proteins. **You can do a lot of other things with the help of these peptides, and it is actually a new way to look at the origin of life. I prepared a series of projects but could not do them in Rome because I had to shut my lab.**

So, I decided the following. That I will publish these projects in the new edition of my book, *The Emergence of Life,* **requested by Cambridge University Press. The idea is to invite readers, possibly researchers in the field, to carry out such projects. It is a way for my ideas to go on experimentally.**

Suzan Mazur: Do you think life is algorithmic or non-algorithmic?

Pier Luigi Luisi: I take a phenomenological approach to this question of the origin of life. Instead of considering a theoretical algorithm and so on, I look at life where life is in its simplest form -- unicellular organisms. There the essence of life is metabolism based on self-maintenance.

Self-maintenance is the essence of life, this is the capability of every organism to maintain itself by regenerating from its inside the very components. I make my own hemoglobin, my own beard, the apple tree makes its own apples and leaves, all from within. **Life is a factory that makes itself from its own borders -- autopoiesis.**

We remain always the same despite the millions of chemical transformations that go on inside us. This is possible because life remakes all those components that are being destroyed by metabolism. We get a little older in the process, but that's another question.

There is a main philosophical point here: the criterion of life does not need to make reference to heredity, genetics, DNA or self-replication. A dog playing in the street is not to be called alive because he is the product of his parents' reproduction! Of course self-replication is the basis for evolution and biodiversity, but it is obvious, looking at the many examples we have around of individual life, including lab artifacts, that you do not need evolution to define life here and now. It is another conceptual blunder of many people working in the RNA camp to identify the criterion of life with the criterion of evolution. Those are two different things.

Suzan Mazur: What are your thoughts about the origin of body form?

Pier Luigi Luisi: I have a systems approach. In my latest book, *The Systems View of Life* [Cambridge University Press], Fritjof Capra and I make the general point that it is the relations among objects that should be emphasized, more than the objects *per se*.

My body is alive because it is an integrated system, where lungs and heart and brain and tissues and nerves, etc. are in continuous dynamic interaction with themselves. Life is coordinated, it's integration. Death is fragmentation, when the organisms and parts do not talk anymore with each other.

This notion is important in relation to the dominant DNA-centric view today, that life is DNA or the genome. I really believe that this view is wrong. It is my entire body as a living

organism which decides how genes should be activated, with a lot of ions, and various factors that make a signal that finally reaches the right point to activate genes to make proteins. This activation, this decision, is a systems property. So we are not our genes, genes are just mechanical soldiers, which are even prisoners. We, living organisms, are the system activating the genes.

I do not know if this is the answer you wanted, or whether your question was more towards ontogeny. Your question also touches on another issue the relation between body and mind, and body and consciousness. We -- Capra and I -- devote many pages to these points in our book.

Suzan Mazur: You did your postdoctoral work in the former Soviet Union. Do you see a difference between the Russian perspective on origin of life and Western perspective?

Pier Luigi Luisi: I was there -- imagine -- 50 years ago, at the time of Nikita Khrushchev. The Soviet Union was really different from today's Russia

The Russians started the origin of life field with Alexander Oparin's book in 1924, *The Origin of Life*. Oparin was the first to invoke prebiotic molecular evolution. This, of course, was in the middle of the Soviet vision of life, and Oparin was influenced by the Marxist dialectic, and by the Darwin book. Those were the two dominant visions of science because they were both without God. Very materialistic in a way.

This is what we accept today, but I am also hastily adding that our modern view is not anymore simply materialistic. You cannot understand the properties of life using the properties of atoms and molecules. But this would be another long story.

After Oparin I do not see any major breakthroughs in Soviet origin of life science. I have not seen any important origin of life experimental work coming from Russia.

Suzan Mazur: Do you think there's still a rivalry of sorts between East and West in science and even between the US and Western

Europe, in spite of the Internet?

Pier Luigi Luisi: Perhaps there is, but it is something of modest proportions. We in Europe, for example, think that journals like *Nature* and *Science* are strongly biased in favor of American authors, and certain American academic addresses in particular. This may be so or not, I do not know.

In terms of philosophical differences between the European and American schools, there is perhaps a difference in terms of style of writing. Americans are much more hyper, sometimes they seem to be directed more toward mass media than toward science colleagues -- see certain big bubbles in synthetic biology "making life" and things of this sort. The European way of writing and presenting papers is more Germanic, more sound and down to earth. But those are generalizations, with their clear limits. . .

Suzan Mazur: How is science funded in Italy? Do the Italian people share in the decision making process about allocation of public funds in science? Do citizens sit on your national science panels, or is it like in the US where the public funds science but the scientists control it?

Pier Luigi Luisi: In a way it's a very simple answer. In Italy since many years, the funds for research have been almost non-extant. After 30 years of working in Zurich at ETH, I came back to Italy, and the difference was shocking.

Think as a matter of comparison: the new professor who took my place at the ETHZ, although I do not know the precise figures, probably received from ETHZ around €1M to start his new lab; he got five assistants paid completely by the Swiss state; he received about €100,000 per year for current expenses; and, of course, access to all Swiss National Fund grants.

When I came to Rome, I received nothing, although my Roman colleagues were very friendly to me -- but there was simply no money. I could do experimental work only because I bought from ETHZ, my old instruments, second-hand (private money) -- and brought them to Rome -- with horrible custom problems (caused by the Italians). And later I could pay my co-workers only thanks to Europeans or international grants.

Suzan Mazur: Do you have a national science organization like the National Science Foundation?

Pier Luigi Luisi: There is something like that but it has very, very little money.

Suzan Mazur: Does the public participate or are there scientists only on the panel?

Pier Luigi Luisi: No. Only scientists. It's peer-review of projects.

Suzan Mazur: Like the United States. What is the extent of corporations and private philanthropy financing scientific research?

Pier Luigi Luisi: Private philanthropy is practically unknown in Italy. Italian millionaires and corporations are blind and deaf to the problems of science.

Suzan Mazur: So you don't have legislation in Italy as we do in the US where we have a virtual "commercial usurpation of the whole scientific enterprise."

Here in the US we have something called the Bayh-Dole Act where a corporatization of universities was allowed to take place with companies acquiring monopoly rights to products developed by scientists funded by the public. So the public is left out of the profit and the information, plus the public has to pay again to use the product following its development.

Pier Luigi Luisi: I am not aware of anything like that. In terms of legislation for research, we are probably still at the times of Galileo.

Suzan Mazur: Corporations don't have partnerships with universities and monopoly rights to products developed by scientists?

Pier Luigi Luisi: There is very little and only for very applicative projects. For basic science you will not find anything like that.

We do have the advantage that there is no interference of non-science groups, but the problem is there is no money. Let me add that in Italy or Germany or France, most of the money for research comes from the European foundation. As I told you, my research in Italy has been supported by international grants coming from the European Union or from the Human Frontier in Japan.

Suzan Mazur: You emphasize wholeness in science. Would you explain?

Pier Luigi Luisi: The basic idea is that everything is connected to everything. In science obviously biology is connected to chemistry, to bio astronomy, to anthropology, to philosophy, and so on. . . Wholeness in science means that it does not make sense to consider biology, for example, as an individual body detached from all other branches. This sometimes is called trans-disciplinarity, but we have to be careful about that. This does not mean that we have to make a cocktail of all those things, and call it wholeness of science.

The main point is still high professionalism in what you are doing -- without watering-down science. Wholeness means that from your own professionalism you have to open your mind to see and accept the points of view of other disciplines, possibly integrating them.

I am often accused to be doing philosophy, and in fact I wrote a few papers for philosophical journals, but I answer by saying that you cannot do science without looking into the philosophical basis of what you are doing.

There is another way to explain the wholeness. It's just like you have a pond and a frog, and the frog in the pond can only see a couple of frogs around itself. But the eagle that flies over the pond has a systems view, sees all the components of the pond interacting with each other. The bird also sees the trees on the edge of the pond and sees that the water in the pond comes from the mountains and pollution from the nearby factory. All possible relationships are considered.

The systems view, in general, says that reality is the product of a series of relations. That is why I gave you earlier the example of the body, of life, as an integrated, interacting system. The same is true for a city, the same for a nation. This is the holistic view.

Piet Hut

Origins of Life and Herding Cats

PIET HUT
(photo, courtesy Piet Hut)

"Science got started in a modern form say with Galileo only 400 years ago. The oldest expressions of art, the oldest expressions of depicting something outside ourselves are in cave paintings of 40,000 years ago. For 99% of human history there was art and religion and for the last 1% there was science. So science often sounds like a teenager, like an adolescent, arrogant and rebellious. That's because it's so young. In another 400 years, if we could speak again, I think science will shape up quite a bit. Get social. . .

Science started by taking an easy path. It focused on the object pole of experience. It did not want to say much about the subject pole. It did not want to say much about the interaction. . . In 400 years we've been able to build up an enormous body of knowledge about objects. . . The unwritten hope in science is that by empirical method taking only one pole, studying that in great detail, you can go back to the back door to the pole door into the rest of the empirical and try and reconstruct everything. But that is at the moment a hypothesis. It may or may not be true. It may not even be clear whether or not one can find a criterion that that is true. . . I think there is room for philosophy and people from other ways of knowing. . .

[T]he current scientific world picture is a picture of only objects. And often when I reflect on that I feel like the person who for the first time went to the moon and looked back at the Earth. . . We forget that . . . what is given here for each of us. . . everything you see is being painted in your consciousness. And that is the first tool you use to study everything else. It is like a blind man who uses a stick to walk around, and with a stick he can feel the whole room. But if you ask what is he really feeling, it is only the stick and the vibrations in the stick. So yes, we have a very good scientific, objective, empirical method, but everything we are experiencing is given in our experience. That is our stick. And in that sense we are blind, and yet we can understand a lot of what is going on around us." -- **Piet Hut**, Modern Cosmology Roundtable

March 22, 2013

Institute for Advanced Study astrophysicist Piet Hut moves lithely, unassumingly, harmoniously through a crowd often in signature t-shirt (unlike Piethut, the asteroid named for him orbiting the sun with a semi-major axis of 2.4 AU, an eccentricity of 0.12 and an inclination of 8 degrees). Hut and I missed one another at the Princeton origins of life conference in January but caught up recently at COOL EDGE 2013, the strategic talks on origin of life at CERN, where we agreed to have a longer conversation when Hut returned home to IAS in Princeton.

Hut, a native of The Netherlands, was the youngest professor given tenure when hired at IAS almost three decades ago, later riding out a lawsuit to dismiss him because of his interdisciplinary interests, which IAS said cut into his responsibilities in the astrophysics department. But colleagues and the media rallied to his defense with IAS dropping the case and naming him Professor of Interdisciplinary Studies.

Hut's work now involves explorations in cognitive science and the philosophy of science as well as computational astrophysics.

As a long-time friend of Japanese science (since 1985), he is also now affiliated as an astrophysicist with the Earth-Life Science Institute at Tokyo Institute of Technology -- Japan's MIT. The next major origin of life symposium takes place there March 27-29. Hut gives one of the opening addresses of the ELSI conference – "The Big Questions" -- and chairs its last session on exoplanets and astrobiology.

Hut says he has a deep appreciation for Japanese literature and culture, as well, and has spent several months a year in Japan for decades.

While he is known for co-writing a simulation algorithm used in measuring movement of stars and helped create the world's fastest supercomputer, in recent years he's been on the rise as a public intellectual, for example, appearing at Manhattan's Rubin Museum in a one-on-one Brainwave 2010 chat with Indian film director and actor Shekhar Kapur: "Does Chaos Have Meaning?" and at other thoughtful gatherings.

He was also one of five physicists participating in the Sixth Mind and Life Conference in India with the Dalai Lama. But the thread that runs throughout his professional career is his talent for bringing other people together around the world to talk to one another. Discussions he's helped organize include subjects such as -- "Fundamental Sources of Unpredictability" (Santa Fe Institute); "Ambiguity Brought into Focus"; "Dialogues with Time and Space"; "Some Unsolved Problems in Astrophysics"; "Deflecting Asteroids" (NASA); to name a few.

Piet Hut's MSc degree in astrophysics is from the University of Utrecht in The Netherlands. His PhD is in particle physics and astrophysics from the University of Amsterdam.

My conversation with Piet Hut follows.

Suzan Mazur: The conference at Tokyo Tech end of March related to origins of life -- directly and indirectly -- is the third major scientific gathering on the subject in three months, with meetings at Princeton and CERN in January and February, respectively. Do you see a ramping up of interest in the field?

Piet Hut: It's hard to say. I'm actually new to the origins of life field. I've been largely involved with physics and other areas of investigation. It was only in December of last year as the Earth-Life Science Institute in Japan started up that I found myself one of the principal investigators, when we decided to delve into the topic of origins of life.

Suzan Mazur: Will the Tokyo conference be streamed over the Internet the way the Princeton conference was?

Piet Hut: I don't think so. They are still setting up the institute's infrastructure, so I'm almost certain that they won't be able to stream the symposium, although I'm almost certain that they will be able to do it for next year's conference. By then they'll also have their own building and staff, and everything should be smoothly up and running.

Suzan Mazur: Do you see new ideas emerging from these origin of life gatherings and a synthesis of sorts taking place, or do you think we are still largely on a fishing expedition regarding origin of life?

Piet Hut: Origins of life is a very broad problem. It will take many decades to make significant progress, although I am surprised how much has been accomplished in the last few years. At the same time, there's such a gap between chemistry and biology, and it will take quite a while to close that gap. And not only between chemistry and biology, but also with geology, the relevance of which is now beginning to be understood to a small extent. That is one of the main threads at ELSI, not only to create a dialogue between chemistry and biology but to bring in geology.

Suzan Mazur: You were a trustee of the John von Neumann

Supercomputer Center for some time, what are your thoughts about algorithmic origins of life?

Piet Hut: Some part of it lends itself to large-scale simulations for sure. So I think it was a very good idea to come to CERN and see some of the enormous computer power they have there, which could be used especially during the period that the beam is down and they are overhauling the hardware.

There are many types of simulations that can be done on the level of pure chemistry, on the level of simple levels for emerging life, on the level of more complex processes within cells. All kind of network studies.

Suzan Mazur: You've been associated with Japanese science for almost three decades. Why?

Piet Hut: My PhD work was in math, particle physics and astrophysics, theoretical things. Higgs calculation was part of my PhD, which made it really nice to be at CERN for the first time.

Following my PhD, I moved to the United States and shortly after I got into heavy computational work, computer simulations in astrophysics. At the time the Japanese were playing a leading role in computing, having very fast supercomputers and there were some very good people doing algorithms. Also, I was culturally interested in Japan. I had always enjoyed reading Japanese literature, playing Go, a Japanese/Chinese board game, and doing all kind of activities relating to Japan. So the combination was a real attraction for me.

In 1984, I visited Japan. I then spent the whole summer there and the entire year of my first sabbatical in 1989. I have been spending a few months every year in Japan ever since.

Suzan Mazur: Do you have a clearer idea since our conversation in Geneva, of how many origin of life researchers there may be out there? Institutions working on the problem? And funding figures?

Piet Hut: In Japan or worldwide?

Suzan Mazur: Worldwide.

Piet Hut: I really don't know -- 500 researchers worldwide, which I mentioned to you, was a very rough number.

Suzan Mazur: Günter von Kiedrowski in a discussion with me in Geneva about "the unthinkable," a subject CERN distanced itself from, said "it is very likely" that we are not alone. Kei Hirose, host of the upcoming origin of life conference at ELSI also told me that he does not think we are alone. What is your view?

Piet Hut: I certainly think that it is fairly likely that there will be other life in the universe both on a very simple level and maybe even on the more multicellular complex level like we have on Earth.

Suzan Mazur: You have several pet projects, Kira Cafe, the Harry's Bar After Hours Chats at Princeton and others. Which of these is most interesting at the moment?

Piet Hut: All interesting, but lately I'm focused on my latest project, which is working with Japan's Earth-Life Science Institute on origins of life. I'm especially interested in the more abstract or meta aspects. I was offered a role at ELSI because the institute is interested in astrophysics and I am an astrophysicist. ELSI is also looking at the geological, geophysical and biochemical aspects of the origin and evolution of life.

Going beyond the Earth with its environment, life on other planets may be very different from ours. There's no particular reason, why, if it exists, it should have RNA and DNA like we have. So ELSI is interested in what you could call universal life. Astrobiology.

What is of great interest to me, and what I will be working on myself, is what are the general principles behind life, not just the biochemistry. What is the complexity inherent in the robustness of life, in the resilience of the way life self-organizes?

Suzan Mazur: Carl Woese was thinking along these lines before he died a few months ago. He was looking at the principles of the origin and evolution of life and told me he did not think that our Last Universal Common Ancestor was anything material, it was a process.

Piet Hut: Yes. I think that is reasonable. It certainly wasn't an individual cell that evolved into life as we know it.

Suzan Mazur: Are you optimistic about where we are headed as a civilization?

Piet Hut: It is so hard to predict. Any moment anything can happen making things a lot worse or a lot better, and the better things have bad side effects, and crises have good side effects. It is completely impossible to predict.

One thing about education that bothers me is that the most interesting insights in physics or in astrophysics -- the kind of academic knowledge I know best -- like quantum mechanics and relativity theory are not being taught in high school. If you have high school physics, you basically have 100-year-old physics. Modern science is not being taught in a way that is understandable.

If I were a high school teacher, I would try to find a way to give some sense of what the core of modern physics is without detailed mathematics, without detailed formal explanations. There are many popular textbooks with quantum mechanics and relativity theory that could be used.

Suzan Mazur: There are initiatives for digital education.

Piet Hut: That will definitely help. Yes.

Suzan Mazur: Are you optimistic about the origin of life experiment taking off at CERN?

Piet Hut: I'm pretty optimistic because the impression I got, more or less as an outsider, is that they have been working on this for two years now and there was enormous good will from both sides -- CERN and the origins of life people. What probably makes the process go a little bit slower is that the culture is so different. It just takes time to speak each other's language.

Of all the organizational models in science, I think CERN is probably the most regimented, army-like -- in a positive sense of the word -- organization where they do experiments. Because they have to be. With 3,000 people working on one experiment, it has to be a hierarchical structure or it will never work.

On the other hand, the origins of life, it's like trying to herd cats, everybody running in their own direction.

So they are really at two ends of the spectrum of organizational discipline. Therefore, the interdisciplinary -- no pun intended -- they face a bigger gap than almost any interdisciplinary collaboration. But I think they will figure it out. They have enough good will.

Suzan Mazur: But where do you see particle physics coming into play with origin of life?

Piet Hut: Not in any direct way. I don't think there is a direct connection with CERN. I think it is the management experience and the computational resources. Those two seem to be the most important. What I hadn't realized was what Günter von Kiedrowski said about building a piece of machinery, some sort of small chemical reactor, and again using the management of CERN. That was the first time I'd heard about it.

Suzan Mazur: Will the camps come together -- the RNA world approach and the metabolism-first model?

Piet Hut: I think these camps are already intermingling and dancing around each other. It's the nature of research that you polarize a little bit to make your positions clear. Then you fight it out and it's revised in the end. It's an intellectual evolution approach.

Suzan Mazur: Would you like to make a final point?

Piet Hut: Yes, young people -- postdocs, graduate students and others -- should organize themselves. In the field of particle

physics, for example, there is little a postdoc or graduate student can do because it's an enormous structure, a matter of navigating like a cog in a machine. But origins of life is still a herding-cat science, everything is still nomadic, so a grad student or postdoc has a chance to make an impact. Carl Sagan did this as a graduate student, he organized international conferences. Got people together from the Soviet Union and America. He was an extremely good creative organizer.

Suzan Mazur: This is what you've spent many years of your life doing as well.

Piet Hut: Yes, I've always enjoyed doing it.

Suzan Mazur: What is the importance of origin of life research at this point? Why does it make a difference?

Piet Hut: There is a philosophical and a practical answer. The philosophical answer is if you look around in the world and try to understand it, you sooner or later come up with three types of questions. There is the question: Why is there anything as all -- matter and energy and space and time? That is more the area of the Big Bang -- how did things get started?

The second big question is, if you look around and see space and time and matter and energy -- why does some of the matter and energy organize into life? Life is such a bewildering thing compared to physics. So how did life get started is the second really big question.

The third really big question is how did self-awareness and intelligence arise, how did individuals appear who could ask the first two questions?

Of those three questions, the one relating to self-awareness and intelligence, which requires brain research, is the most complex and least understood. The Big Bang is roughly understood, not in detail, but the basic way of asking the question is clear. Origins of life is the middle question, still a big gap and a big question mark between chemistry and biology, but it's not hopeless.

The timing is right. The reason that I chose particle physics as part of my PhD 20 or 30 years ago was because black holes, the Hawking mechanism, Big Bang, etc. were just getting to an accelerated stage and there were many new discoveries. This is still an interesting topic, but 30 years ago origins of life was still a very, very tough topic and now we are in an accelerated phase. **If I were a graduate student today, I'd probably choose origins of life over physics.**

Suzan Mazur: Is there an urgency to the origin of life investigation?

Piet Hut: I would say opportunity, not urgency. Scientists are opportunists, they look under the light pole because if you look away from the light pole, you can look pretty hard but you won't find much. Fortunately, the light is shifting now. This is an opportune time for studying origins of life.

In another generation it could be brain studies, neuroscience. The whole question of how first-person subjective knowledge and third-person objective knowledge are two sides of the coin in the brain, as electrical and chemical processes, and at the same time seem to be the seat of our subjective experience. Discovering how those two connect I would say is another generation from now.

So somehow we are in the middle of about a 100-year period in which generation by generation, one by one, the three big questions come to the fore. We are very lucky. That is the philosophical part.

The origins of life is the second of the three big questions and the practical part is that for many applications in biology if the origin of something is known, this helps in understanding related questions. Practical applications will sooner or later result from the investigation of origins of life.

Denis Noble

Replace the Modern Synthesis (neo-Darwinism)

DENIS NOBLE
(*photo, courtesy Denis Noble*)

May 9, 2014

In a search for the more colorful side of physiologist and systems biologist Denis Noble, I was drawn to his Oxford Trobadors page of Occitan music (medieval songs of love and chivalry from the south of France, Italy and Catalonia), featuring videos of the group mixing it up with modern infusions of jazz, etc. Noble, a classical guitarist, doubles as troubadour and maestro in the clips -- with impressive stage presence. He says, "No one needs to be just a scientist."

Denis Noble's understanding of music is clearly reflected in the elegance with which he communicates science. Maybe that's partly why he was invited to China to talk about evolution and the need to move beyond neo-Darwinism. (Noble prefers the term "modern synthesis.") He reports in our interview (below) that "youngish" scientists came up to him after his address to the Chinese Association of Physiological Sciences describing their struggles trying to get published in Western journals.

Denis Noble is Emeritus Professor of Cardiovascular Physiology and Co-Director of Computational Physiology at Balliol College, Oxford University. He also serves as president of the International Union of Physiological Sciences (he was IUPS Secretary-General for almost a decade) and is currently editor-in-chief of the Royal Society's bimonthly journal *Interface Focus*, which features articles at the crossroads of the physical and life sciences. Noble is the author of *The Music of Life* and 10 other books as well as 500 scientific papers.

It was Denis Noble's discovery, more than 50 years ago at University College London, of the "electrical mechanisms in the proteins and cells that generate the rhythm of the heart" -- the basis of his PhD thesis -- and his mathematical model that first attracted international attention to his work. He and his colleagues at Oxford are now making computer simulations for the rest of the organs of the body.

Noble has received several honorary PhDs and numerous other awards, including the British Cardiovascular Society's Mackenzie Prize, the Russian Academy of Sciences' Pavlov Medal, the Pierre Rijlant Prize from the Belgian Royal Academy of Medicine, the British Heart Foundation Gold Medal and the Baby Medal from the Royal College of Physicians in London. He is a fellow of the Royal Society and an honorary member of both the American and Japanese Physiological Society.

My interview with Denis Noble follows.

Suzan Mazur: In recent years the modern synthesis has been declared extended by major evolutionary thinkers (*e.g.*, "the Altenberg 16" and others), as well as dead by major evolutionary thinkers, the late Lynn Margulis and Francisco Ayala among them. Ditto for the public discourse on the Internet. My understanding is that you are now calling for the modern synthesis to be replaced.

Denis Noble: I would say that it needs replacing. Yes.

The reasons I think we're talking about replacement rather than extension are several. The first is that the exclusion of any form of acquired characteristics being inherited was a central feature of the modern synthesis. In other words, to exclude any form of inheritance that was non-Mendelian, that was Lamarckian-like, was an essential part of the modern synthesis. What we are now discovering is that there are mechanisms by which some acquired characteristics can be inherited, and inherited robustly. So it's a bit odd to describe adding something like that to the synthesis (*i.e.*, extending the synthesis). A more honest statement is that the synthesis needs to be replaced.

By "replacement" I don't mean to say that the mechanism of random change followed by selection does not exist as a possible mechanism. But it becomes one mechanism amongst many others, and those mechanisms must interact. So my argument for saying this is a matter of replacement rather than extension is simply that it was a direct intention of those who formulated the modern synthesis to exclude the inheritance of acquired characteristics. That would be my first and perhaps the main reason for saying we're talking about replacement rather than extension.

The second reason is a much more conceptual issue. I think that as a gene-centric view of evolution, the modern synthesis has got causality in biology wrong. Genes, after all, if they're defined as DNA sequences, are purely passive. DNA on its own does absolutely nothing until activated by the rest of the system through transcription factors, markers of one kind or another, interactions with the proteins. So on its own, DNA is not a cause in an active sense. I think it is better described as a passive data base which is used by the organism to enable it to make the proteins that it requires.

The third is an experimental reason. The experimental evidence now exists for various forms and various mechanisms by which an acquired characteristic can be transmitted.

So I think the reasons for replacing the modern synthesis are the experimental, that certain forms of inheritance of acquired characteristics have now been both demonstrated and their mechanism worked out, and the more philosophical point about the nature of causality. I believe that the modern synthesis, and indeed very many aspects of the interpretation of molecular biology generally, got the question of causality in biological systems muddled up.

Suzan Mazur: Lynn Margulis told me the following in 2009:

> "[W]hat Haldane, Fisher, Sewell Wright, Hardy, Weinberg *et al.* did was invent. . . . The Anglophone tradition was taught. I was taught, and so were my contemporaries, and so were the younger scientists. Evolution was defined as "changes in gene frequencies in natural populations." The accumulation of genetic mutations was touted to be enough to change one species to another. . . . No, it wasn't dishonesty. I think it was wish fulfillment and social momentum. Assumptions, made but not verified, were taught as fact".

Margulis added that "people are always more loyal to their tribal group than to any abstract notion of truth. Scientists especially tend to be loyal to the tribe instead of the truth."

Would you comment?

Denis Noble: I would certainly go along with the view that gradual mutation followed by selection has not, as a matter of fact, been demonstrated to be necessarily a cause of speciation. Many of those who defend the modern synthesis would say, "Well, it has been." But what you find when you look at the examples modern synthesists give is that they are for the gradual transition of one species into another in the historical record.

Just to take an example of that: the so-called ring warbler example. With the ring warbler you can watch the process, work back the historical process of how these birds developed into various subspecies around the southern areas of the world, south of the Himalayas and eventually creeping around to meet again at the north of the Himalayas. What you find is that each of the varieties can breed with each other all the way through the various branches that lead from the south to east and west. When they meet in the north, they no longer interbreed. What that tells us is there clearly was a historical development in which these warblers developed first into subspecies and then eventually into different species in the sense that they don't any longer interbreed. That tells us that that process of speciation occurred, but it does not tell us the mechanism by which speciation occurred.

Regarding wish fulfillment, what I find is that the modern synthesists tend to quote such ring warbler examples as though it is obvious that they must have occurred by gradual mutation followed by selection, when it isn't certain that that can be the mechanism if other mechanisms exist. You have to prove it. So I go along with the view that there has been no really clear proof that speciation occurred via gradual mutation followed by selection.

Suzan Mazur: Do you think scientists are anywhere near an agreed-upon definition of what life is?

Denis Noble: That's an enormous question. No, I don't think so, actually. First of all, molecules are dead. There's no sense in which individual molecules can be said to exhibit the phenomenon of life. You need a process among many, many components -- molecular components, of course, included -- of an organism in order to have something which has some of the characteristics we would want to regard as living.

You can produce a list, and physiologists do this a lot, of the obvious characteristics of a living organism: It grows, it divides, it reproduces, it metabolizes. But one can find examples where some of those properties would not necessarily be present in all the examples that one would want to take of a living organism. That's why we have difficulties with organisms like viruses, which clearly can't reproduce on their own. Are they or are they not living? It's obviously a very difficult question to ask.

I'm not sure that we need to bother about a precise definition. It's pretty clear that DNA on its own, proteins on their own, metabolites on their own, lipids on their own, are not alive. It's the network of system interaction that can be said to have living characteristics, however one defines those.

Suzan Mazur: University of Chicago microbiologist Jim Shapiro, whose work you cite, told me in our 2012 interview that he no longer uses the word "gene," saying:

> "[I]t's misleading. There was a time when we were studying the rules of Mendelian heredity when it could be useful, but that time was almost a hundred years ago now. The way I like to think of cells and genomes is that there are no "units"; there are just systems all the way down."

New York Medical College cell biologist Stuart Newman said he thinks the gene is "down but not out."

But only a week or so ago the science section of *the New York Times* ran a piece touting "de novo genes" and their appearance and disappearance.

What is the status now of the gene in your view?

Denis Noble: First of all, I go along largely with Jim Shapiro's view of the difficulty of the definition of a gene. I think it's actually even more difficult than Jim says. My argument is very simple. Wilhelm Johannsen in 1909 introduced the definition of "gene." He was the first person to use that word, although he was introducing a concept that existed ever since Mendel. What he was actually referring to was a phenotype trait, not a piece of DNA. He didn't know about DNA in those days. We now define a gene, when we attempt to define it, as a particular sequence with "start" and "stop" codons, etc., in a strip of DNA. My point is that the first definition of a gene -- Johansen's definition as a trait, as an inheritable phenotype -- was necessarily the cause of a phenotype, because that's how it was defined. It was, if you like, a catch-all definition of a gene. Anything that contributed to that particular trait -- inheritable, according to Mendelian laws -- would be the gene, whether it is a piece of DNA or some other aspect of the functioning of the cell. That we define "gene" as a sequence of DNA becomes an empirical question, not a conceptual necessity. It becomes an empirical question whether that particular strip of DNA has a function within the phenotype. Some do and some don't.

It's interesting that many knockout experiments don't actually reveal the function of the knocked-out gene. In yeast, for example, there's a study that 80 percent of knockouts don't have an obvious phenotypic effect until you stress the organism. What that tells me is that we have progressively moved from a definition of a gene which made it a conceptual necessity that the defined object was the cause of the phenotype -- that's how it was defined -- to a matter which is an empirical discovery to be made, which is whether a particular sequence of DNA plays a functional role or not. Those are very, very different definitions of a gene.

So I go further than Jim. Not only is it difficult, as he says in his book, to now define what a gene is; one should be thinking more of networks of interactions than single and fatalistic genes at the DNA level. It's also true that the concept of a gene has changed in a very subtle way, and in a way that makes a big difference to how the concept of a gene should be used in evolutionary biology.

The reason for that is very simple. It is that many of the definitions used by modern synthesists, including Richard Dawkins, are actually the Johannsen definition of a gene -- that is, the trait as the phenotypic characteristic.

[**Note**: While many consider Dawkins a Darwinist rather than a neo-Darwinist, Noble says that neo-Darwinists tend to encourage people to think they are following Darwin when, in fact, they're not. Noble points out that Darwin did not exclude the inheritance of acquired characteristics: "He even praised Larmarck, in the preface to the fourth edition of *The Origin of Species*."]

Suzan Mazur: There's also natural selection, which became a catch-all term. As Richard Lewontin has pointed out, it was intended as a metaphor not to be taken literally by generations of scientists. The range of views about what natural selection is is staggering -- a brand, a political term, a political and scientific term, failure to reach biotic potential, physicists are seeing it as part of a larger process now, etc. etc. Things are being majorly redefined.

Denis Noble: You're putting your finger on a very important point here. And what I just said about the definition of a gene is only one example where I think some philosophical clarity is needed.

Suzan Mazur: Is it the case that there are all sorts of mechanisms at play, some of which have now been identified, that have been previously considered part of natural selection? It seems natural selection is used as a catch-all for a failure to identify what the mechanisms are.

Denis Noble: I think that's right. In principle, Darwin didn't refer to any mechanisms. It was simply what we now regard as a fairly obvious statement, which is if there is variation and no definition -- not in Darwin's books, anyway -- as to what the cause of that variation might be, if there is variation, then there can be selection. If there can be selection on variants, then some will survive and some won't. In some sense this is a necessary truth, isn't it?

Suzan Mazur: Congratulations on your lecture on evolution at the Chinese Association of Physiological Sciences. What was the response to your talk there in China?

Denis Noble: Very interesting. I had similar responses in India. A lot of scientists came up to me afterwards and said, "Thank goodness somebody from either Oxford or Harvard said what you said." I asked them, "What do you mean by that? I mean, it shouldn't matter where somebody comes from whether what they're saying is correct or not." They said they were working on various epigenetic forms of inheritance, in some cases on cross-species cloning between different species.

Suzan Mazur: Are you referring to scientists in China or India?

Denis Noble: Actually, I'm thinking of both, but China is certainly where the cross-species cloning work occurred. It was India where the epigenetic work was occurring. And the comment was, "We can't get our papers published." They said they simply have been told that their theory is wrong.

Suzan Mazur: This is fascinating because Lynn Margulis mentioned to me just a few years ago that the Chinese lack any tradition in evolution, although they enjoy "superb" traditional healing medicine. It was wonderful that you gave this address in Suzhou. Seems like a real breakthrough.

Denis Noble: I do think there is a different overall approach, which has a lot to do with the integrative aspects of traditional Chinese medicine. But that is also true of India. Ayurveda, which is the Indian equivalent of traditional Chinese medicine, has more or less similar kinds of characteristics -- that is, emphasis on the integrative whole, how the past contributes to the whole, but also understanding the whole. So I'm not too surprised by the reactions in both China and India.

Suzan Mazur: Can you tell me who some of the key evolutionary thinkers are in China?

Denis Noble: Now, there I don't really know. The people who came up to me were youngish scientists telling me that they tried to get their work published and had great difficulty finding places to publish it.

Suzan Mazur: You said some of the papers focused on hybridization.

Denis Noble: Yes, that I do know. That is the work of Yong Hua Sun and his colleagues at Wuhan at the Chinese Fish Institute. I referred to their work in my lecture. What Yong Hua Sun *et al.* did was take the nucleus of one species of fish and insert it into the denucleated but fertilized egg cell of a different species. What they got as an adult -- it's very rare that you get an adult from such a cross-species clone -- but what they got as an adult is intermediate between the two, whereas, of course, in a gene-centric view you should -- and assuming the genes are defined as DNA -- you should get the animal from which the nucleus was taken. That doesn't happen in Yong Hua Sun's experiment.

Suzan Mazur: Where was Yong Hua Sun trying to get published and was rejected?

Denis Noble: In the end they got their work published in a journal called the *Biology of Reproduction*. That's not an evolutionary journal. I think they were fortunate in having an outlet, because, after all, this could be said to have to do with reproduction and mechanisms of reproduction -- the mixed nucleus and egg, nucleus from sperm and egg from the mother. It was fairly obvious to them to publish their work in *Biology of Reproduction*, which they did six or seven years ago. There's an article by Yong Hua Sun in the *Journal of Physiology* at the end of this month.

Suzan Mazur: Are you aware of any other cutting-edge evolutionary research that China is doing aside from the work on hybridization?

Denis Noble: No.

Suzan Mazur: China is clearly doing interesting work on stem cells.

Denis Noble: Unfortunately, I didn't take notes in China or India on who approached me with their stories of difficulties in getting things published.

Suzan Mazur: Are the Chinese investing in origin of life research, including protocell development?

Denis Noble: I would be very surprised if they are not, but I don't know for certain whether that is the case.

Suzan Mazur: What is your view as to whether or not we are alone in the universe?

Denis Noble: I find it very hard to imagine that we are alone. There's absolutely no reason why the conditions that enabled life and living systems to emerge on Earth shouldn't occur in many other places in the universe, given its immensity and the rapidity with which we're finding that systems similar to the Solar System must exist in many other star systems. It would be utterly remarkable if we were alone in that sense.

Now, whether life on many other locations -- planets, if you like -- necessarily evolved into being what we are is a rather different question, of course, because remember we're the last few seconds, if you think of it in terms of a speeded-up process where a billion years becomes a year or something like that. We humans are just the last few seconds of life on Earth, and there's no guarantee that anywhere else there's going to be something that is really at all similar to us. I say "no guarantee," but that doesn't say that it couldn't exist. What I suppose I'm saying is that it is extremely unlikely that we are alone in the sense that the evolution of life only occurred on Earth. I think that's most unlikely. It must have occurred elsewhere.

[**Note**: Denis Noble and University of Chicago microbiologist James Shapiro have recently teamed up to create a web site: www.thethirdwayofevolution.com featuring dozens of major scientists (the list grows) who have issues with neo-Darwinism.]

James Shapiro

The Evolution Paradigm Shift

JAMES SHAPIRO
(*photo, courtesy James Shapiro*)

"Given the exemplary status of biological evolution, we can anticipate that a paradigm shift in our understanding of that subject will have repercussions far outside the life sciences".
-- **James A. Shapiro**, *Evolution: A View from the 21st Century*

May 7, 2012

I called University of Chicago microbiologist James Shapiro, who's now also blogging on *HuffPost* about science, to arrange an interview after noticing that we'd both recently been bashed by Darwinist Jerry Coyne in the same column. I reached Shapiro at home. He was engaging, although he described himself as a "reclusive person" -- which he says he finds key to serious thinking. The commotion was over Shapiro's book: *Evolution: A View from the 21st Century*, since Coyne, also a University of Chicago professor, has an evolution text he'd like to keep relevant. I decided to have a look at Shapiro's book and see exactly why Coyne was agitated.

Shapiro was traveling soon, so I scrambled to pick up a copy of the book. Curiously, none of the Manhattan book stores had one and neither did the libraries. Columbia University's only two reference copies were out being read. But the New York Institute of Technology, situated in the middle of a forest on Long Island, did have one and provided the perfect canopy to immerse myself in Shapiro's world of cognitive cells.

Shapiro says the key science journals have yet to review the book, including *Nature* and *Science*, but it's being discussed continuously on *HuffPost* where Shapiro has encouraged a rigorous debate on topics like "Jerry Coyne Fails to Understand Yet Again" and "What Is the Best Way to Deal with Supernaturalists in Science and Evolution?".

He graciously takes on some of the blogosphere's notorious "Wild West" commentators as well, who frequently mistake him for an anti-evolutionist.

The University of Chicago Laboratory Schools was where James Shapiro began his scientific investigation -- the same grade school that inspired Lynn Margulis. His BA is from Harvard, *magna cum laude*, in English Literature, which gives him a certain edge on the Internet.

Shapiro's PhD is from Corpus Christi College, Cambridge, in Genetics, where he was a Marshall Scholar. He did postdoctoral work at both the Institut Pasteur in Paris and at Harvard Medical School.

Curious about alternative forms of government, he moved to Cuba in the early 1970s where he taught genetics at the University of Havana after leaving the research lab at Harvard Medical School concerned about the potential misuse of his work on genetic

396

engineering of *E. coli*. He returned to the US for another postdoc at Brandeis two years later. In 1973, Shapiro joined the faculty at the University of Chicago where he currently teaches in the Department of Biochemistry and Molecular Biology.

James Shapiro has served on the editorial board of a half dozen science journals, including the *Journal of Bacteriology* and *Enzyme and Microbial Technology*. He is a Fellow of the American Association for the Advancement of Science as well as the American Academy of Microbiology and a member of numerous other professional organizations, among them, the American Society for Microbiology and American Society for Biochemistry and Molecular Biology. In 2001, he was made an Honorary Officer of the Order of the British Empire for his service to the Marshall Scholarship program -- having served many years as its Chicago regional chairman -- which enables American students to study at some of the finest schools in the UK.

Shapiro has also been honored with a Darwin Prize Visiting Professorship at the University of Edinburgh. He was Visiting Professor at Tel Aviv University and Visiting Fellow at Churchill College, Cambridge, among other distinctions.

His books include: *Evolution: A View from the 21st Century* (FT Press), *Mobile Genetics Elements* (Academic Press), and with Martin Dworkin (eds.) *Bacteria as Multicellular Organisms* (Oxford University Press).

James Shapiro was first to propose a detailed mechanism for replicative transposition as a means of DNA mobility in genomes. His research also encompassed pattern formation in bacterial colonies. In 1984, observing that bacterial colonies resembled flowers in organization, not unlike the patterns Nobel Laureate Barbara McClintock saw in maize kernels, Shapiro later concluded that "bacterial colonies could also be viewed as multicellular organisms."

My interview with James Shapiro follows.

Suzan Mazur: What first sparked your interest in evolutionary science -- you have an undergraduate degree in English literature from Harvard and a PhD in genetics from Cambridge.

James Shapiro: I became interested in biology as an undergraduate. The topic of evolution just kept coming up. As a research student at Cambridge, when I began to focus on mutagenesis, evolution was again right there because mutation was the source of the raw material of evolution. My first big lesson in evolutionary science was that the mutations I was studying in bacteria were unexpected and unpredicted. People had actually missed them because they accepted the current version of mutations just as point mutations. Here was something quite different. Pieces of DNA inserting themselves in the genome.

Later, I found unexpectedly that starvation triggers a big increase in DNA rearrangements. I also observed some genome changes occurring in patterns in bacterial colonies. All of that gave me a lively interest in evolutionary subjects.

Suzan Mazur: Do you come from a science family?

James Shapiro: No, not at all. My father was a businessman and my mother a homemaker. I was interested in lots of things growing up. I had a great science teacher in grade school -- a fairly well known woman named Bertha Morris Parker at the University of Chicago Laboratory Schools I attended -- who I'm sure must have had some influence on me.

Suzan Mazur: But were you always interested in nature?

James Shapiro: No, I think I was interested in figuring out things, making sense of things. I didn't collect insects and stuff like that.

Suzan Mazur: What took you to Cuba in the early 1970s where you taught for a couple of years? And what did you teach?

James Shapiro: I taught genetics. What took me there was the fact that everybody was talking about alternative forms of government, about revolution, about social change, and so forth. And here was a chance to see it up close.

Suzan Mazur: What was the conclusion that you drew from

398

seeing it up close? Did you have a favorable experience?

James Shapiro: Well the experience had its ups and it had its downs. I don't want to go into too many details, but I would say the take-home lesson was that human nature trumps everything. Any system has its strong points and its weak points.

Let me explain what I mean about human nature. At a benefit we were introducing the new director of an opera company where I'm on the board. He was asked to explain the season that he was planning. And he said: "Well, it's about power and love and love of power."

You can see those motivations in every society.

When the society works well, as ours has, you're fortunate. We're going through a period where we're not so fortunate because political dialogue has broken down, much like the dialogue about evolution and science education.

Autocratic regimes sometimes can be positive. I was talking to a woman about Rwanda where Kagame is not a democrat but at least he's trying to advance the nation. So the Rwandans are a lot luckier than the Congolese, for example.

So the human factor dominating every political system was the take-home lesson.

Suzan Mazur: You've now got a pretty lively blog on *Huffington Post* -- how does this square with your description of yourself as "reclusive"?

James Shapiro: I think I am a fairly reclusive person. I think that's necessary to do serious thinking.

Suzan Mazur: But how does that square with your lively blog on *HuffPost*?

James Shapiro: The blog has to do with trying to advocate a certain way of thinking about evolution and life in general. This was an opportunity presented to me by *Huffington Post*. They asked me if I'd like to blog. They get people to blog for free and that gives them content. And it gives those of us who have something to say the position to argue, a platform. Having published the book, I felt it was worth taking the argument forward. Also, I'm using it as a kind of test bed for learning how to write for a more general audience.

Suzan Mazur: You're reaching out in substantive ways and getting some quality responses. Not as much of the usual rabble.

James Shapiro: Well, you do get abuse and obsession in the blog comments. On Friday, I put a note up saying that the blog was getting rather *ad hominem* and away from the substance. Since then the messiness hasn't completely stopped but the comments have been somewhat more substantive. It's the Wild West in the blogosphere.

Suzan Mazur: Statements have been attributed to you regarding a crisis in evolutionary science, and you expressed that in your 1997 *Boston Review* article on a "Third Way."

[**Note**: James Shapiro and Oxford's Denis Noble have recently teamed up at: www.thethirdwayofevolution.com]

Are you saying that for at least a half century or so vast billions of dollars of public money has gone to scientists whose work has been rooted in a belief system -- *i.e.*, the metaphor of natural selection, as Richard Lewontin described it in the pages of the *New York Review of Books*?

James Shapiro: I would say research based on theories that will be superseded is inevitable. I was quite struck when I read Thomas Kuhn who understood that. I was sitting by a swimming pool in the Dominican Republic at a meeting on plasmids and he was writing about 18th century chemistry and physics. As I was reading I was saying to myself -- "Wow, that's the way biology operates today."

Kuhn captured something very quintessentially human about the scientific enterprise: that you inevitably never capture nature as it is. You only capture a portion of it that you can

400

figure out and theorize about. And you go on exploring that portion of nature. For some period of time the explorations are extremely productive. But over time and as technology develops, partly as a consequence of what the scientific enterprise is doing, new phenomena come up and can't be explained any longer in the same way. In the end there are always a group of people who defend the existing belief system more than is justified by the empirical observations.

Tension arises between those who say the empirical observations are telling us something different and those who defend the intellectual framework which led to those empirical observations. I am not immune to being unable to appreciate where new approaches can lead. For example, I was one of the people who initially thought genome sequencing was just an excuse for using technology without any idea of what we were going to find. I believed that people had run out of useful ideas for experimental biology and were doing DNA sequencing as a substitute. I was totally wrong about that. It turned out that sequencing has been extraordinarily revealing and far from a waste of time. No matter what kind of ideas lay behind it, it's opened up a treasure trove of new ways of thinking about genomes and DNA in evolution.

So the answer to your question about the money is that money is always being spent based on ideas which are ultimately going to prove fallible. As I put it in a blog, if Newton couldn't get it right, what hope is there for the rest of us? But it's not a waste of time and money as long as the research is based on real empirical science, because the observations then lead to a more sophisticated way of thinking about things.

Suzan Mazur: With so much money being blown on war and other shenanigans and with Paul Krugman today on PBS confirming that we are now in a DEPRESSION -- are you really saying that public money should continue to go to scientists, America's most successful intellectuals in securing public funds, who agree to work to ensure the country's health, national security and economic stability, even if they're publicly defending a Darwinian system they won't privately defend to other scientists? How much of the scientific establishment is doing this, by the way?

James Shapiro: I couldn't answer that question because I only know the part of the scientific establishment -- and not that much of it -- that I've been exposed to in the course of my professional life. And the last 10 to 15 years have been fairly reclusive. I faced a major decision in 1984. I could spend all my time on the road or I could stay home and get some work done. I decided it was better to stay home than to spend my time on the road. It wasn't necessarily the best decision professionally but it was a decision that suited me at the time.

The way I look at it, most of my colleagues exhibit a schizophrenic attitude towards new ways of thinking. Somebody will be doing research on a subject like the role of reversed transcribed DNA inserted into the genome and creation of new genetic loci. The shorthand for this is neogene formation. And that's a totally non-Darwinian process. Certainly nothing that any of the Darwinians could ever have conceived or predicted.

The experimentalists do the research and they talk about it in appropriate new ways. In practical terms they are talking in a totally different way about how genetic change happens than fits the conventional wisdom.

However, if you say to them you're doing non-Darwinian science, they say, "No, what are you talking about? We're doing Darwinian science." Because, in their minds, "Darwinian science" is synonymous with non-supernatural scientific exploration of evolution and genome change in biology.

I don't think there's a conscious deception going on. I describe it as a kind of schizophrenia.

Suzan Mazur: You're saying there's not a conscious deception going on. That's interesting.

James Shapiro: Let me just add to that.

Naturally there are very few people who will come out as I have and say: Look the emperor has no clothes. People are not willing to do that. They probably are very wise not to. But somebody has to say it. I'm in a position to do that and so I say it.

Suzan Mazur: But if a collapse in this philosophical belief system has occurred -- as you've said: "the DNA record definitely does not support the slow accumulation of random gradual changes transmitted by restricted patterns of vertical descent"--

James Shapiro: Yes, that's a statement of empirical observation.

Suzan Mazur: So is science now without an acceptable explanation as to how evolution happened?

James Shapiro: No I don't think so. We see bits and pieces of the whole process. Certainly we have paleontological evidence. We have the comparative biology. It started off as comparative anatomy but it's gone much farther than that, of course. All of this tells us about relationships. And now we have the genome evidence, which solidifies our view in the evolutionary relationships. It complicates the picture, but it adds an element -- which is the one I've been focusing on -- the process of genome change itself that is critical. That is what I call "natural genetic engineering."

Suzan Mazur: In pinpointing some of the most obnoxious behavior in defense of Darwinian scenarios, I am reminded of the keynote speaker of the Rockefeller University Evolution symposium -- University of Chicago biologist Jerry Coyne -- who stood before an audience of distinguished scientists in the spring of 2008 to do damage control, first trashing creationism and then declaring that he could cite 300 examples of natural selection but didn't have enough time to do so. The speech was arranged by the National Center for Science Education. . . .

Are Coyne and his pal Richard Dawkins, by not publicly recognizing that a sea change has occurred, milking lucrative performances and book deals? And if so, isn't this a disservice to science?

James Shapiro: Well that's a loaded question.

I've gone on record in my blog as saying that I thought that Dawkins and his ill-conceived atheist crusade hurts science education and hurts evolutionary science. I've criticized Jerry Coyne for making statements that he can't support. Both Coyne and Dawkins have been characterized as having a neo-atheist agenda in attacking religion. And it is true that many of the comments on my blog reflect a deep reservoir of anti-religious sentiment. I've noted in the blog that we should get out of the business of attacking supernaturalists. It's not in science's interest to make a war with religion or religious belief.

Suzan Mazur: And get on to the science.

James Shapiro: And get on with the science. And also to have a little bit of humility. Science never provides ultimate answers.

Both Coyne and Dawkins are doing well for themselves by being vitriolic and vehement in their campaign against religious belief and against some of the more foolish things that religious fundamentalists do. But I suspect ego gratification is the major driver more than financial gain. Obviously it's nice if you make money at the same time. . . .

Suzan Mazur: At the moment, politics is a large part of evolutionary science.

James Shapiro: I would like to see, as much as possible, an emphasis on rigorous scientific debate. We should not let all these philosophical or political issues -- who controls the funds, etc. -- dominate. It's counterproductive. There are real serious open evolutionary scientific issues to discuss and there are real serious alternative ways of looking at them. It's extremely important that science move forward in new and unexpected directions.

Suzan Mazur: Your book was published by FT Press, Pearson, parent of the *Financial Times* and *The Economist*. I find it fascinating that the Brits are standing behind an American scientist who is challenging Darwinian science, one of their national treasures. . . . I went through the book over the weekend. It's very thoughtful the way you've put it together. Would you describe your theory, which involves cells speaking to one another

-- cognitively, informationally. You say in reality the "gene" is "not a definite entity" -- it's "hypothetical in nature."

James Shapiro: There are three components there.

(1) As I say in the book, cells do not act blindly. We know from physiology and biochemistry and molecular biology that cells are full of receptors. They monitor what goes on outside. They monitor what goes on inside. And they're continually taking in that information and using it to adjust their actions, their biochemistry, their metabolism, the cell cycle, etc., so that things come out right. That's why I use the word cognitive to apply to cells, meaning they do things based on knowledge of what's happening around them and inside of them. Without that knowledge and the systems to use that knowledge they couldn't proliferate and survive as efficiently as they do.

(2) We've learned a great deal about hereditary variation through molecular genetics studies. I was personally involved in this back in the late 60s and 70s and since then we've learned about a wide variety of biochemical systems that cells use to restructure their genomes as an active process. Genome change is not the result of accidents. If you have accidents and they're not fixed, the cells die. It's in the course of fixing damage or responding to damage or responding to other inputs -- in the case I studied, it was starvation -- that cells turn on the systems they have for restructuring their genomes. So what we have is something different from accidents and mistakes as a source of genetic change. We have what I call "natural genetic engineering." Cells are acting on their own genomes in a large variety of well-defined non-random ways to bring about change.

This is consistent with what Barbara McClintock first discovered in the 30s when she was studying chromosome repair and then later in the 40s when her experiments uncovered transposable elements. All of these natural genetic engineering systems are regulated or sensitive to biological inputs. That sensitivity is what we've learned about cell regulation in general. As I say, cells don't act blindly, and they don't act blindly when they change their genomes.

(3) So if genetic change is not a series of accidents and not a series of necessarily small changes, then how does it work out in evolution? That's where the DNA record from genome sequencing comes in and confirms what many of us had argued for a long time: namely, all of these systems of genetic change, of natural genetic engineering, have played a major role in evolutionary change. We have a new view of how cells operate in evolution, which is much more information technology friendly.

I think the first blog I put out was quoting a December 2011 paper where they went through the human genome using the 29 mammalian genomes that had recently been aligned. The authors concluded that, at a minimum, there were 280,000 different components, defined functional elements in the genome that came from mobile genetic elements.

The point is that natural genetic engineering systems have played major roles in evolutionary change. We also see in the DNA record that evolutionary change has not just been a slow accumulation of random changes.

A good way of summarizing this is to compare the genome to storage systems in computers. The conventional view is that the genome is a read-only memory (ROM) system that changes only by copying errors. Incorporating what we have learned at the biochemical level about the cellular and molecular processes of DNA change, we can formulate a fundamentally different view. The contemporary idea is that the genome is a read-write (RW) storage system that changes by direct cell activity. How cell control circuits guide that change activity is the scientific issue of the moment.

Suzan Mazur: So what the gene is, how it first appeared and when, are an old way of thinking about things.

James Shapiro: The gene first appeared at the beginning of the 20th century with the rediscovery of Mendelism. Gregor Mendel called them factors, which is fine because it's nondescript. Then Wilhelm Johannsen came up with the term "gene." And over time the gene became endowed with a whole bunch of properties.

There's a 1948 *Scientific American* article by George Beadle in which he called the gene the basic unit of life.

Suzan Mazur: I mean in evolutionary time. This thinking that the gene arrived at some point in the emergence of life. It seems to be an old way of thinking now because the definition of the gene has become much more ambiguous.

James Shapiro: When three scientists rediscovered Mendelism at the turn of the century, in 1900, breeders started seeing discrete hereditary differences that could be passed on from generation to generation. And so the idea that you could have a particulate or atomistic view of the genotype built up, and then the individual components were called genes.

We now have a more sophisticated understanding of hereditary. You've got an integrated, super-sophisticated storage system called the genome. You can't just try and reduce it to any one of its components.

I don't use the word "gene" because it's misleading. There was a time when we were studying the rules of Mendelian heredity when it could be useful, but that time was almost a hundred years ago now.

The way I like to think of cells and genomes is that there are no "units." There are just systems all the way down. This idea came to me unexpectedly in conversation during a visit to give a lecture at Michigan State. A colleague said that his goal was to discover the basic units in the genome. Without thinking about it consciously, I responded, "What if there are no units?" At that moment, I realized that this answer was something I had been thinking about for a long time.

There have been lots of surprises and lots of discoveries along the way to a systems view of the genome: coding sequences being broken up into exons and introns, non-coding sequences which serve as signals for expression of coding sequences, different ways of reading the coding sequences, and so forth. When you have all of that complexity in genome expression, you no longer can give any kind of simple unitary definition of what you mean by a particular piece of the genome.

With George Beadle and Edward Tatum in the 1940s, you had the one gene-one enzyme hypothesis. It was thought that we could say definitively that the business of "genes" is to determine the structure of proteins. But now we have all of this so-called "noncoding" information in the genome. In our own human genomes, "non-coding" sequences greatly exceeds the protein coding capacity. A lot of that "non-coding" DNA is clearly functional and very important for genome action. So we're beginning to develop a far more sophisticated idea of what a genome is and how it operates. That's all a part of bringing evolutionary science into the 21st century.

Suzan Mazur: But how far back in time would you say were cells talking to one another without genetic systems, *i.e.*, programs?

James Shapiro: I think I make it explicit in the book that we don't have enough knowledge yet of how cells came into being in the first place.

Suzan Mazur: When do you anticipate that might become more clear?

James Shapiro: We need to understand how the cells that exist today operate. That's going to require another shift in our thinking because we have a very mechanical, again a very atomistic view of that.

We don't yet understand how cells and organisms are integrated functionally and informationally. When we understand that integration, then we'll have a better idea than we do right now of what the basic requirements are for life and for reproduction.

I expect there will also be technological changes in paleochemistry aiding the search for traces of early life. We don't have this right now. It's possible we may never have it. On the other hand, science always amazes us with what it's able to find. I don't want to be in a position to say we can't work something out scientifically because very often we do succeed in unexpected ways.

Suzan Mazur: When did multicelluarity first happen?

James Shapiro: At the first cell division. Life for as long as we know it has been multicellular. The single celled organism is -- not exclusively, but by and large -- a synthetic construct devised partly to analyze how cells operate and partly as a consequence of Koch's postulates and the germ theory of disease. In studying bacterial pathogenesis, the emphasis was on isolating a pure culture from a single cell. But in nature very few cells exist isolated from other cells.

Suzan Mazur: When do you think evolution began? How do you think about it?

James Shapiro: This is part of what I think is a new understanding of what it takes to be alive. I would include the ability to change as a fundamental feature of living organisms, as a basic vital function.

Suzan Mazur: Are we including pre-biotic evolution?

James Shapiro: There are people who want to speculate about pre-biotic evolution. I don't think we can talk about it in a serious scientific way.

Suzan Mazur: Interesting.

James Shapiro: I think we need to come to terms with the biology that exists in front of us before we're able to speculate about what might have preceded it. And I think we're very far from being finished with that enterprise.

Suzan Mazur: Should amateurs who are seriously challenging Darwinian scenarios be welcome to the evolutionary science discourse?

James Shapiro: Well I don't see how you can exclude anybody. The point is not who's saying something or what their credentials are but the value and substance of what they're saying. Having started off my career as an amateur with a degree in English rather than in biology, I think sometimes that's an advantage because you are without prejudices. You're freer to understand and interpret data.

I wasn't the first person to see mutations caused by transposable elements in bacteria but for some reason I was the first one to be able to free myself of thinking that there had to be point mutations, base substitutions or frame shifts, which was the reigning theory of mutation at the time.

There's even a paper by Crick and Brenner and colleagues called "The Theory of Mutagenesis" that was published shortly before my own work which said that base changes and frameshifts were the substance of genetic change.

I was able to say, maybe my mutations are additions of DNA, and it turned out to be correct. That's how we started to study transposable elements in bacteria. People working on antibiotic resistance soon found that antibiotic resistances hopped around. So yes, people can come in from all kinds of backgrounds. It's the substance of what they're observing and saying that matters.

Suzan Mazur: Would you wrap up your view of 21st Century evolution and where we're headed?

James Shapiro: We have the three components, which are:

(1) Cells act in what I call a cognitive way or an information processing way. Some people like to say "computational." The only reason that I don't use the word computational is that it doesn't include the sensory aspect of how cells operate. And the sensing and its molecular bases are all very firmly established scientifically. There's no question about it.

What we don't understand is how everything is integrated, how the information is processed and how the cells end up doing the appropriate thing. We know a lot about the components involved in signal transfer and decision-making, but we don't know how the whole system works. That I think is the key frontier in the 21st century. The research will not only impact biology, but it will possibly revolutionize computation as well.

(2) Cells engineer their own genomes and they do it in a wide variety of ways that are subject to sensory inputs and which can be targeted within the genome. I document that pretty extensively in the book.

(3) We know from the DNA record that natural genetic engineering systems have been important in the evolution of new life forms.

The key questions that I see in evolution science besides learning more about those three components are:

(i) What is the link between ecological change and genome change in organisms?

(ii) What is it about the natural genetic engineering processes and how they are regulated and controlled that biases them towards creating new functionalities?

We know we can stimulate rapid genome change in the laboratory by starving cells, or putting them under pressure or in high salt and other stress conditions. Similarly, by manipulating their genomes the way McClintock did so they don't operate normally. Or by hybridizing, as in horticulture, having different species mate or different populations mate. All of those things will trigger very significant episodes of genome restructuring. And we know genome restructuring has played a role in evolution and evolution is marked by the appearance of biological functional innovations.

Suzan Mazur: But do you think scientists are now moving onto the same page around the world regarding evolution or do you think European science is heading in a different direction than American science, for example?

James Shapiro: European science -- and this was one of the virtues of doing a PhD in England and a postdoc in France -- European scientists have always taken a more intellectual view of what they do. American science is very operational and highly technological and much less conceptual than science in Europe. I think sometimes there is more openness and sometimes more hostility towards new ideas in the States.

Suzan Mazur: But ultimately do you think there will be a coming together and that the public will be served by these different approaches.

James Shapiro: I think we have to realize that there are a whole range of new players coming into the science game.

Suzan Mazur: Right.

James Shapiro: The Japanese have always been very important. It was the Japanese who first figured out transmissible antibiotic resistance, which solved one of the big evolutionary events that we've observed happening in real time, which is the emergence of multiple antibiotic resistance in bacteria.

Suzan Mazur: And then the Chinese coming in. Important research centers in the Middle East.

James Shapiro: The Singaporese, Russians and Indians. All of that's going to be good -- scientists coming from different cultural and philosophical traditions who are not bound by the history of European or American science.

Suzan Mazur: So an Anglo-American approach to evolutionary science is no longer enough, it's got to be more inclusive.

James Shapiro: That's always been the history of science. . . . There's a large period when the French dominated science. And the Germans and German-speaking dominated. Johannes Kepler was a German. Carl Linnaeus was a Swede. Nicolaus Copernicus a Pole. Pre-Renaissance science was largely carried out by Chinese and Arab researchers.

Suzan Mazur: But it's Darwinian evolutionary science that's taken root. And that's Anglophile.

James Shapiro: Yes, there is an aspect of nationalism there that's gotten bound up in the Darwin -- Lamarck rivalry. Whether that's positive or negative I don't really know.

I think we're headed to a more diverse scientific culture because people from these countries have their own traditions. The Chinese think differently about things than we do. They like to make lists, for example.

Suzan Mazur: So it's going to be a while before there's a real coherence to evolutionary science.

James Shapiro: Well I suppose. People are always going to try to unify, and other people are going to be diversifying. The unifications are always going to involve trying to impose artificial constraints on natural phenomena, and I think they're doomed to failure.

We have this terrible dilemma in science. We need to be reductionists to get meaningful results and make observations. But when we take the observations and try to understand what they mean, then we have to stop being reductionists and become integrationists to understand how the things we've identified and singled out fit into the whole picture.

We've lost sight of that need for integration with the successes of molecular biology. But I think we're getting back to an integrationist view now because people are studying complex problems like cell biology and multicellular development using molecular tools. It's becoming clear that there's an interaction between the parts and the whole which is far more complex and multidirectional than people used to think.

I think that that shift from reductionism to integrationism actually needs to happen in the physical sciences as well. They still hang on very much to the idea that you can have "a theory of everything." I'm rather dubious about that.

Again, my experience in science has taught me that you should never say that something can't happen because we're continually discovering things we've been told can't happen. It's just been a few years since we've realized cells can pick up fragments of sequence from invading DNA. Whether they do it at the DNA level or the RNA level is not entirely clear yet. But they can pick up fragments of a sequence from invaders, incorporate them into their genomes and then defend themselves. We were told that was

413

impossible based on the Luria-Delbrueck experiment.

Being more inclusive and more open to new ideas and imaginative approaches will serve us best. It should be difficult for new ideas to become accepted because that's how you really test their worth, their value. But it does not make any sense from a truly scientific point of view to exclude things *a priori*.

Suzan Mazur: People should be encouraged to put their theories out there.

James Shapiro: Yes. It's an Anglo-Saxon idea, the marketplace of ideas. People who try to erect artificial barriers are making a big mistake.

Part 7

It's the People's Circus

"I insist that Dr. Schekman speak to me directly about the quality of the science, that, in the end, you are trying to protect. . . ."
-- **Lynn Margulis**

"I make very little distinction between the scientific community and the corporate community. Those are very close links."
-- **David F. Noble**

Chapter 16

"Marvelous" Lynn Margulis

LYNN MARGULIS
(*photo, courtesy James MacAllister*)

"Grants are awarded by your colleagues who sit in Research Councils and Foundations. Most of us, in any establishment, tend to be conservative and to follow what is called the paradigm. This creates a cycle of submission. . . . The disregard for science's ethical principles is widespread." -- **Antonio Lima-de-Faria**, Professor of Molecular Cytogenetics, Emeritus, Lund University

In Memory of Lynn Margulis, 1938 – 2011.

January 5, 2010

D oes a science peer review system based on secret submission policies benefit the American public who fund science? A review by this author of correspondence between the prestigious *Proceedings of the National Academy of Sciences of the United States of America* -- the print weekly and online daily research journal (paid subscription) of the National Academy of Sciences -- and the authors of several recent scientific papers, most eventually published by *PNAS*, reveals a nasty back story about submission procedures that in some cases work against the best interests of the public as well as sound science.

The uproar had to do with three papers submitted to *PNAS* several months ago by NAS member Lynn Margulis, a recipient of the US Presidential Medal for Science. One of them, "Destruction of spirochete *Borrelia burgdorferi* round-body propagules (RBs) by the antibiotic Tigecycline," the authors say involves an excellent candidate antibiotic for possible cure of the tick-borne chronic spirochete infection Lyme Disease in the US, recognized as "*erythema migrans*" in Europe and elsewhere. However, the paper was held up because *PNAS* said it had issues about the way Margulis chose her reviewers on the first (unrelated) paper she presented, that is, Donald Williamson's "Caterpillars evolved from onychophorans by hybridogenesis." As a result, all three papers were stuck. The last of the three, also on spirochetes, which Margulis says was properly and favorably reviewed, has not yet been approved for publication. . .

Margulis is one of 2,100 US members of the NAS. She does not receive government funding and has further distinguished herself by refusing to take DARPA (Defense Advanced Research Projects Agency) money. Margulis admits she is viewed by some within the NAS as "contentious" but says she "only wants to see that real science, open to those who want to participate, is well done, discussed critically without secrecy and properly communicated."

NAS promotes itself as a private, non-profit organization of distinguished scientists that serves the "general welfare," although it was actually incorporated in 1863 by Congress during the incoln presidency with a mandate to further the investigation of and report on science and art whenever called upon by any department of government. And in 1884, it was authorized *"to receive and hold trust funds for the promotion of science, and for other purposes"* (emphasis added). NAS is joined at the hip to the National Research Council, National Academy of Engineering and Institute of Medicine -- collectively called, the National Academies.

When a reporter for *Times Higher Education* in London asked *PNAS* why the Margulis-introduced papers were on hold, she was told: "The submission process is confidential and we cannot comment on any papers currently under consideration." In reality, a paper on advice to the editor by anonymous expert reviewers can be pulled at any point on its way to *PNAS* publication.

One of the reviewers Margulis selected for the Williamson paper reported receiving an intimidating call from an editor at *Nature* magazine and commented, "It sounded like he was trying to discredit the work and that I might have been a weak link."

Margulis then learned that one of these anonymous expert *PNAS* reviewers was blocking publication of the Williamson paper, although she suspected that this was just the tip of the iceberg. She wrote me to say that she was always surprised that they ever had let her into NAS (elected in 1983). She wondered what the qualifications of the anonymous reviewer were and noted that the delay "was cruel . . . when so many people are suffering Lyme arthritis and this Brorson microbiology paper (that I actually co-author) provides a fine clue to eventual adequate treatment . . ."

Margulis who had just returned to her teaching position and lab at the University of Massachusetts-Amherst from Oxford University, where she'd spent the 2008-2009 academic year as Eastman Professor, then advised *PNAS* managing editor Daniel Salsbury of her course of action:

> "If [*PNAS* editor-in-chief] Randy Schekman or you or anyone else at the *PNAS* continues to pit Williamson's, Robert Higgins's, Professor Mark McMenamin's, Oxford

418

Professor Martin Brasier's and my authority about marine larval evolutionary history against an anonymous expert reviewer and refuses to be satisfied with my reviewing procedure and therefore to block the entirely unrelated Brorson *et al.* paper, I am going to be forced to request a signed legal statement that Randy Schekman, you and the anonymous outraged reviewer in fact have more authoritative knowledge than we do about these evolutionary lineages. I humbly request that you do not force me into this position as I am not a litigious person. . . . The *PNAS* arguments are from authority and procedure and not from science. . . . I insist that Dr. Schekman speak to me directly about the quality of the science, that, in the end, you are trying to protect. . . ."

While Margulis won the *PNAS* battle on both the Williamson and Lyme Disease papers, and one outraged critic came to light in a *PNAS*-published letter -- Harvard University's Gonzalo Giribet -- questions remain about just what kind of science is promulgated at *PNAS*. (Giribet told me by phone he was not a "reviewer" on the original Williamson paper but also that he would never admit to being an "anonymous reviewer" if he were an anonymous reviewer.)

Another NAS member (foreign) has been in the process of suing *PNAS* for the same sort of last-minute expert board member rejection over a paper of his on nitrite in the water supply and cancer victims in China.

Harvard scientist Richard Lewontin, considered by many to be the "most important evolutionary biologist of the passing generation," resigned from NAS, describing the Academy as a "political organization which is almost quasi-governmental" and "not about to refuse the DOD and military establishment" (2003 interview with Harry Kreisler of the Institute of International Studies at the University of California, Berkeley). Lewontin left NAS because its operating arm, the National Research Council -- funded by federal agencies -- had committees that were doing secret war research.

We have no way of knowing if anything has changed regarding secret war research. In 2008, US government agency grants and contracts to the National Academies totaled $192.3M with unspecified funding from private and nonfederal sources at $52.7M. Some of those federal grants and contracts came from myriad branches of the Defense Department, Office of the Director of National Intelligence, National Security Agency and the Executive Office of the President (George W. Bush).

[**Note**: *PNAS* has now ended its fast-track submission process.]

David F. Noble

Peer Review, Where Are The Scholars?

DAVID F. NOBLE
(photo, courtesy Sarah Dopp)

In memory of David F. Noble, 1945 – 2010.

No, passing peer review is not the scientific equivalent of the *Good Housekeeping* seal of approval. . . I confess that decades ago as a Hearst Magazines fledgling I would on occasion pass by, and with some curiosity, peek in the glassed-in *Good Housekeeping* Institute where the coveted seal of approval was given to mattresses able to withstand umpteen thrusts, to pantyhose, pots and other utilitarian items. And how can I forget running into Johnny Weissmuller in the hall one day? The seal meant something -- even Tarzan wanted to be associated. But to help me flesh-out what scientific peer review IS exactly, I decided to contact the Tarzan of science and technology historians, David F. Noble.

David Noble, at the time of our interview, was a tenured professor in the Department of Social and Political Thought at York University in Toronto. His activism about matters involving the politics of science first surfaced in the early 1970s when he said he was "coerced" by the University of Rochester to sign away rights to his doctoral thesis in history in order to get his PhD.

Noble's undergraduate degree was in history and chemistry from the University of Florida. On and off for 10 years he worked as a biochemist at Tufts, Purdue and the University of Rochester Medical School.

Noble went on to teach at MIT in the Science, Technology and Society department, beginning in 1978; he was denied tenure in 1984, fired "for his ideas and his actions in support of those ideas." Noble's ideas were at odds with the escalating corporatization of universities and MIT was a university with tight corporate ties. He said the reason he was denied tenure was political and described MIT's tenure process as "*ad ho*c."

MIT Institute Professor Noam Chomsky commented at the time that Noble was "too radical for MIT." Nevertheless, Noble sued MIT for $1.5M, and although he later dropped the suit, he did manage to get the court to make public the case documents and order MIT to review its tenure practices. The American Historical Society rallied to Noble's side, condemning MIT's decision to dismiss him.

Noble moved next to the Smithsonian, where as curator of industrial automation he was fired again, this time for including the only extant hammer used by the Luddites in the 19th century -- which also exposed his Luddite leanings.

(Noble as late as 2010, the year he died, did not communicate by email.)

At Drexel Institute he was tenured but chose to snip ties after five years, taking a position at York University where he spent almost 20 years -- although not without controversy.

Noble was, in fact, a champion of social justice throughout his career. Acts of courage (partial list) included: co-founding the National Coalition for Universities in the Public Interest with Ralph Nader in 1983 to address the nightmare of universities selling out to corporations; attention to the issue of science and misogyny; and challenging religious bias in academia.

Regarding the latter, Noble, a Jew, protested York University's policy of canceling classes on Jewish holidays by canceling his own classes on Muslim holidays. He argued that to be fair, classes for all religions should really be canceled on respective holidays if Jewish holidays were observed. In 2004, Noble took the matter to the Ontario Human Rights Commission, and York University was forced to get rid of its policy.

A second action involved Noble's grievance over reprisals from the first action, *i.e.*, his academic reputation was damaged because he had challenged York University over its religious bias. In 2008, he was awarded $2,500 as a settlement.

In a third action, still before the court at the time of our interview, Noble said he suspected "money laundering" by various pro-Israel lobbying organizations. He also said that if he won the $19M lawsuit, he would feel that he'd "made a difference."

And then there was the lawsuit Noble brought and won regarding his being denied an endowed chair at Simon Fraser University, established in the name of the founder of Canada's Social Democratic party. An email was uncovered from SFU's president to vice president saying, "I'd avoid this appointment like the plague."

The humanities department at SFU and the Canadian Association of University Teachers thought Noble was THE GUY for the job and stood by him, with CAUT attesting that Noble's "academic freedom had been violated." Noble sued for "damages for loss of income, past and future; damages for loss of reputation;

423

aggravated damages and punitive damages and costs." In 2007, he received a public apology from SFU plus an undisclosed sum of money.

David F. Noble was the author of the following books: *America by Design*; *Forces of Production*; *Smash Machines, Not People*; *Digital Diploma Mills*; *A World Without Women*; *Progress Without People*; *The Religion of Technology*; and *Beyond the Promised Land*.

I last spoke with David Noble by phone in February 2010, just months before he died at age 65. It was a life too brief for a man who gave so much.

Suzan Mazur: In recent articles I've been focusing on abuse inside the peer review system, which has spiraled out of control -- to the extent that at the low end we now find virtual death squads on Internet blogs out to destroy scientists who have novel theories. They pretend to be battling creationism but their real job is to censor the free flow of ideas on behalf of the science establishment. The science establishment rewards bloody deeds like these by putting the chief assassin on the cover of *The Humanist* magazine, for example.

But you've written in "Regression on the Left" that the problem IS the peer review system itself. Why do you think so?

David Noble: When you say THE problem is the peer review system -- the peer review system in my view is doing what it was designed to do -- censor. And filter. Peer review is a system of prior censorship, prior review -- prior meaning prior to publication. So the idea of abusing the peer review system sort of adds insult to injury, because the peer review system itself is injurious.

Suzan Mazur: The first scientific journal was published in 1665 by the Royal Society of London. Would you briefly timeline events from there, highlighting what the original plan was for a postwar National Science Foundation?

David Noble: My understanding is that peer review in its modern form is very recent, dating from World War II. The participants in what we now call science were people with resources, usually independent wealth. When we go back to the Royal Society we're talking about a very small community of people.

What happens with World War II, for the first time, the taxpayer is underwriting the bulk of scientific research. That never happened before. During WWII -- this is in the United States -- the government set up an operation called the Office of Scientific Research and Development (OSRD), headed by Vannevar Bush, vice president of MIT. The government did this to essentially enlist the support of non-governmental institutions, such as universities -- private universities as well as private corporations -- to do research and development for the war effort.

The government's own research laboratories were clearly inadequate for that task. So what they did was magnify the scope of the state's research and development efforts by bringing in these extra-governmental institutions.

The government invented the contract system. They contracted with private actors, private institutions to do the work for the state for the war

Suzan Mazur: Prior to WWII scientists worked within the military and then after the war the government agreed to finance scientists working out of their own labs.

David Noble: Right. Before the war most research was funded by private foundations. And there were many people in the scientific community. . . **I make very little distinction between the scientific community and the corporate community. Those are very close links.**

Vannevar Bush, for example, was the vice president of MIT but he was also a director of AT&T, Raytheon, etc. Bush *et al*. didn't want the government involved in funding of research because they understood correctly that if the government was involved, then the government, and through the government, the taxpayer would have some say about what was done with that money, what the research agenda would look like, etc. There was real resistance throughout the 1920s, 1930s to the state funding research precisely for that reason.

Suzan Mazur: You already had the National Academy of Sciences.

David Noble: In terms of the scale of operations, that was very small. WWII was the watershed. What happened was the invention of the contract system by the OSRD and there was almost immediately controversy over who was getting grants for these contracts. Since the makeup of this committee was pretty much representative of the elite universities and also the largest corporations, they got the lion share of contract work.

Many outside that charmed circle said this was unequal, a privileged allocation of resources, etc. University of Nebraska said: Why don't we get some of this money, why does it all go to Harvard and Stanford and Chicago? And small companies who were not in that charmed circle made the same complaints. This controversy raged during WWII.

The second question had to do with patents, and I'm not going off on a tangent here. This is all related. What the companies did -- and this is extortion -- they said yes, we will build the tanks for the government but we want the patents on the work we develop. And if you don't give us the patents, then we're not going to do it. And so the OSRD say fine, we'll give you the patents. That was another very, very contentious issue.

Again, they were getting patents, monopoly rights to products developed by the taxpayer. So all of this is raging.

By about 1943-1944, there was discussion about what the postwar scientific establishment would look like. By this time, the corporations and the universities and the scientists who had been reluctant to take federal funds for fear of taxpayer involvement were now so enamored of the largess that they didn't want to give it up. And they said, we can't go backwards -- this is the new game -- we are going to be taking taxpayer money. But we don't want the taxpayer involved in what we do. This is a fundamental challenge to the whole. . .

426

Suzan Mazur: And this is what exists today.

David Noble: Yes, but what happened first is that Harley Kilgore, a senator from West Virginia, set up a plan for a "National Science Foundation" whereby the taxpayer -- an ordinary citizen, a non-scientist -- would sit on committees and panels overseeing the allocation of research funds.

In response to that, Vannevar Bush and his friends put together a counterproposal calling for a "National Research Foundation" -- which became more or less what we have in today's National Science Foundation.

The Vannevar Bush *et al.* legislation said essentially that science would be funded by the taxpayer but controlled by scientists. Again, scientists -- this is important to emphasize -- are not simply scientists, but scientists and the corporations they work for.

Suzan Mazur: Also, to get this government funding scientists have to agree to do a few things for the government. Ensure national security and the health and economic security of the nation. There's a *quid pro quo*. It's a way of further silencing scientists.

David Noble: Well it's certainly been used that way. What I'm saying is something simpler. There was a problem with the way the committees and panels overseeing the allocation of research funds would be set up. The problem had a name and the name is DEMOCRACY. The fundamental tenet of the democratic system is that the taxpayers funding something have control over what's done with the money.

Harry Truman said it was the most undemocratic piece of legislation he'd ever seen and vetoed it. It went through minor changes and became what we have today -- a scientific establishment run by scientists with very little political oversight. The key thing is how they kept the taxpayer out was through PEER REVIEW.

Suzan Mazur: It wasn't until the 1960s that science journals were

427

turned into a money-making racket when the federal government gave its blessing to a per-page fee for publication. Scientists were not only writing articles for the journals for free but now there was a per-page fee for publication. Then the number of science journals increased leading to competition between journals and to mediocre science.

Are you saying such a system should be scrapped? Should the peer review journal system be scrapped?

David Noble: I think so. Oh yeah.

Suzan Mazur: Should anything replace it?

David Noble: If the taxpayer is paying for the research, why shouldn't there be a representative array of citizens on all scientific panels? And again, going back to your first question, the purpose of peer review is prior censorship and I believe very strongly that if people want to criticize something that you write or I write, they have every right to do that AFTER it's published not before it's published. To me that's the critical issue.

Suzan Mazur: People have a right to have their views out there.

David Noble: Right. Now what happened with peer review that began within this establishment to keep democracy out then became the gold standard for everything. So, for example, I've published seven books, I've published countless articles, but I've never ever nor will I ever publish in a peer-reviewed journal or a peer-reviewed publication ever.

According to my university I have no publications. This is literally true. The only thing now that is reviewed are peer-reviewed publications. And the fact that none of my work is peer reviewed -- it might be considered, but secondarily.

Suzan Mazur: Do you have tenure? Does that make you ineligible for tenure?

David Noble: No, fortunately for me I'm close to the exit. I'm a tenured full professor. But that's the situation now in every field.

Suzan Mazur: Has it affected your salary?

David Noble: No, no, no. What I'm saying is for the people coming through now, this is the game.

Suzan Mazur: Isn't it strange that so many busy scientists write these staggeringly complex journal papers for free, that they pay for their articles to be published and thousands of dollars extra if they want the public to read them, even though it's the public who funds the research? And that journals continue to operate in secrecy about revenues and operating costs. . . .

David Noble: What I'm saying is these commercial interests have insinuated themselves into the professional evaluation system. Why do people publish? They publish primarily to promote their own careers. Because that's the measure.

Most of what's published shouldn't even be published. It's published for that purpose. And very few people read it.

Suzan Mazur: Very few people can read it.

David Noble: Or have any interest in reading it. Let me go back to something really simple. When I got my PhD in 1974 from the University of Rochester, I was required to give my dissertation to University Microfilms -- a private company that processes and sells dissertations of theses.

Suzan Mazur: Did you have to pay for this?

David Noble: No. I didn't have to pay, but in order to get a PhD from the University of Rochester I had to sign a commercial contract with University Microfilms giving them my dissertation, and I received nothing from them. And they were going to sell it.

Suzan Mazur: Incredible.

David Noble: So I said to the University of Rochester, I'm not going to sign the contract. And the University of Rochester said to me -- then you cannot get your degree.

Suzan Mazur: Incredible.

David Noble: This was in 1973 and 1974.

Suzan Mazur: And how did that play out?

David Noble: What we're talking about is the insinuation of a commercial enterprise into the credentialing process. It turned out that that was standard in every institution of higher education in the United States.

Suzan Mazur: You got the PhD.

David Noble: Anyone who has gone through that gauntlet -- you're done, you're finished with the thesis. The margins are correct. You've got the right font. You're done. You're out of there. And they say: **"NOW SIGN THIS."**

And I said, "NO. I'M NOT GOING TO SIGN IT."

And they said, "IF YOU DON'T SIGN IT, YOU DON'T GET THE DEGREE."

Well, that's called coercion. So I signed the contract.

Pressure was brought to bear on University Microfilms that they should at least be forced to give royalties. And that did come to pass.

I've never gotten any royalties, but the point is, that's the way they finessed it. The university said this is an aid to us because of bibliographical control -- that this private company is doing this. **A private company had insinuated itself in the credentialing process. And that's what you're talking about.**

Suzan Mazur: Isn't even stranger that bankers run the boards of these big scientific publishing companies . . . . [D]on't you see that as a real problem, where the board of directors of these major scientific publishers making billions of dollars -- publicly-traded companies -- are bankers?

David Noble: What people would say is, well, these commercial

430

enterprises publish our work. And the reason they get away with it is because they've implicated themselves, insinuated themselves into the credentialing and professional evaluation process. So that every academic, every scientist has to publish.

Suzan Mazur: But these bankers don't know the content and they're passing it along to the public. That's the problem.

David Noble: It's one of the problems. . . . What I would say to you is, the way they [scientists] get paid is in their career advancement. That's what we're talking about. The insinuation of these commercial interests into that process. That's how it works. . . .

Let's put it in a larger context -- 90% of research done in universities, even private universities -- is paid by the taxpayer. Any company can come to the university, lay down some lunch money and leverage that 90% and in exchange get contractual obligations for prior review before publication. Non-disclosure agreements. Licensing rights. That's the whole system.

Suzan Mazur: It's all for sale.

David Noble: IT'S ALL FOR SALE.

In 1980, the Birch Bayh-Robert Dole amendment to the Patent Act was passed. This is another watershed. The Bayh-Dole amendment laid to rest the controversy that began in WWII over the patenting of publicly-funded research. Up until 1980, it remained ambiguous.

What the Bayh-Dole amendment said was that the universities automatically now own all patent rights on publicly-funded research. What that meant was that universities were now in the patent-holding business and they could license private industry and in that way give them the rights over the results of the research funded by the taxpayer. It was the biggest give-away in American history.

Most scientists work for universities. Universities from 1980

on established intellectual property policies which said that they universities as institutions, as corporations, own the research done by employees. This is what private companies had been doing since the turn of the 20th century.

You're a researcher at a university. You do some research and you decide that there's something in there that's of value -- the university owns it. And you can maybe cut a deal with the university. The university wants to sell the rights and maybe they'll give you a little piece of it.

If a private company has put in some money to that research effort, they have a contract that gives them rights. One of the rights is PRE-PUBLICATION REVIEW. They want to see the research, check it out for anything that's commercially viable and they can censor. They can say, we don't want you saying anything about that. THIS IS ROUTINE. THIS IS THE STATE-OF-THE-ART RIGHT NOW.

Suzan Mazur: Incredible. Is there a way of breaking up the "academic mafias" so that discovery and the free flow of ideas prevails and individual researchers have a shot at being funded and having their ideas taken seriously?

David Noble: Well, you'd have to revoke the Bayh-Dole amendment.

One stop shopping on this is a book by Jennifer Washburn called *Universities Inc.* -- it's a good chronology of this whole sea change with Bayh-Dole, etc. This is an important part of the story. From every angle, the journals, University Microfilms, all of the commercialization of research – from every angle there's commercial usurpation of the whole scientific enterprise.

Suzan Mazur: You've said "[I]t is perhaps time for the Left once again to put science in perspective." That the Left criticizing the informed critics of science as participating in "anti-science" is a sign that the Left really needs to "return to the revolution." Would you comment further?

David Noble: What I mean there is, and this is what I outline in that article, the Left grows out of a critique of religion in the beginning of the 19th and end of the 18th century. And science was the substitute. They substituted science for religion.

Suzan Mazur: You also say those roots are intertwined with misogyny.

David Noble: Ok. That's another issue. Let's keep it simple. The point here is that science became like God. But since WWII, in part because of Hiroshima and other events, other products of science, critique of science became a very serious matter. And the Left was very much involved in looking anew at looking at science as political. And scientists as human beings and as people with interests, etc. So they de-mythologized science.

It went by many different names. Social construction of science, whatever. For decades people were, and still in some quarters are, looking very critically at this whole enterprise. And then along comes this global warming campaign. And you have these people like George Monbiot and others acting as if there had never been any critical examination of science.

Al Gore -- his whole theme is propaganda. **A consensus of scientists. Well, when you have a consensus of scientists, that should set off alarms. That scientists shouldn't be consensual. There should be all sorts of controversy in science.**

Suzan Mazur: You've also got scientists in evolutionary biology who pound on the creationists because they don't have fresh discoveries themselves. What they're doing is making an industry out of bashing the creationists -- instead of improving the science. That's what's happening on the science blogs, where you get these virtual death squads opposing any science that veers from Darwin orthodoxy. Characters purporting to be atheist scientists who are actually violent Darwin religious cultists censoring the free flow of ideas. Making statements like, "I'm always happy to see a fellow hang himself."

That's the peer review that's now popular. It's degenerated into a bloody massacre.

David Noble: Tribalism is rampant. The idea that people still hold is that science is this community of inquirers and that they review one another's work has never been true. It's always been mythical.
. . .

The peer review thing, the reason why it works is because people's careers are implicated in it. Anyone who wants to be promoted or get a job has to SUBMIT to this regime. I never did. But I'm the exception. And I come out of a different moment in time perhaps. There's no way I could probably get a PhD today. There's certainly no way I could have become an academic. No way. That's what's going on now. People might have concerns about this, but they have no choice but to SUBMIT.

That's what they're told**. So those anonymous peer reviewers have absolute decisive power over people's professional lives.**

Suzan Mazur: So you're saying that one way we can change this is to get the public onto the National Science Foundation and government science panels.

David Noble: Yes.

My criticism of peer review, which for me is no big deal, turns out to be unique. Nobody's talking about this. When George Monbiot attacks Alexander Cockburn by saying that the stuff Cockburn is referring to was not peer reviewed, and I say what kind of an idiot is Monbiot.

David Dixon was a writer for *Science* and *Nature*. . . **There was a time when there were science journalists who were alert to this and understood the politics of science. . . . It's like the Left just went to sleep.**

Suzan Mazur: People have been bought off.

David Noble: Right. And when Alex Cockburn and I and Denis Rancourt raised questions about it, we were just pilloried by the Left, which is mindboggling.

Suzan Mazur: It continues. The attack by the so-called Left regarding the questioning of science and peer review. They say this is the system we have -- you think we're going to throw away thousands and thousands of papers and start from scratch?

David Noble: Well the whole thing's corrupt. . . . Peer review makes things so easy for people who are evaluating. All they have to say is that's a peer-reviewed article. End of story. Gold. They don't have to actually read the stuff.

So, suppose I'm up for promotion and the dean says, oh look he's got six peer-reviewed publications. Well they've been peer-reviewed, I don't have to read them. You see what I'm saying? Takes them off the hook.

Suzan Mazur: It's frightening, very dangerous.

David Noble: Let me tell you this, you'll be amused. York University is the third-largest university in Canada. This past year they amalgamated a number of different faculties into The Faculty of Liberal Arts and Professional Studies, which is the biggest academic faculty in Canada. And they hired this guy with much fanfare from outside to become the new dean.

The president announced his appointment with a lot of trumpets. And he's described as a renowned scholar of Chinese history. Well, I said that's interesting. I wonder who this guy is. And I start looking. A renowned scholar of Chinese history -- let's see what he's written. I couldn't find anything. And I said, this is strange.

I happen to know some real historians of China. I called them up and asked, have you ever heard of this guy? They said no. The long and the short of it is -- the guy is virtually unpublished in the field of Chinese history. He's unknown in the field.

So a group of faculty took out an anonymous Gmail account called York Faculty Concerned About the Future of York University and exposed the fraud, accused the president of fraud. Here's what happened -- NOTHING.

He was made the dean anyway without faculty opposition. Meanwhile, the university hired outside lawyers to sue Google,

Bell and Rogers, the main Internet Providers, to try to get the identity of the people who did the exposé. That's what happened.

But the most important thing is there was no outcry. That was it. The most egregious academic fraud I'd ever experienced. Why was there no outcry? It was all confirmed. The university said well it was a mistake by the media relations people, acknowledging that the guy had no publications. He's still the dean. He's the guy who's reviewing everybody else's CVs and dossiers for promotion and hiring.

Suzan Mazur: Did they put him in charge because he'd be compliant?

David Noble: I think there are other reasons, but the point is he has no scholarly credentials.

Suzan Mazur: Incredible.

David Noble: It is incredible. But the point is he's the dean. That's it. Why was there no uproar? And here's the reason. Because all the faculty have no scholarly credibility either. There's no other explanation. They're all playing these games.

WHERE ARE THE REAL SCHOLARS? There aren't any. Everyone's either broken or bought. There's complete conformity within the community.

Acknowledgments

I am grateful to Peter Sheesley for his cover design and charming caricatures, and to the following people for supporting my work: Alastair Thompson, Anne Russell, David Freeman, Jeffrey St. Clair and the late Alexander Cockburn.

www.ingramcontent.com/pod-product-compliance
Lightning Source LLC
Chambersburg PA
CBHW021545210326
41599CB00010B/314